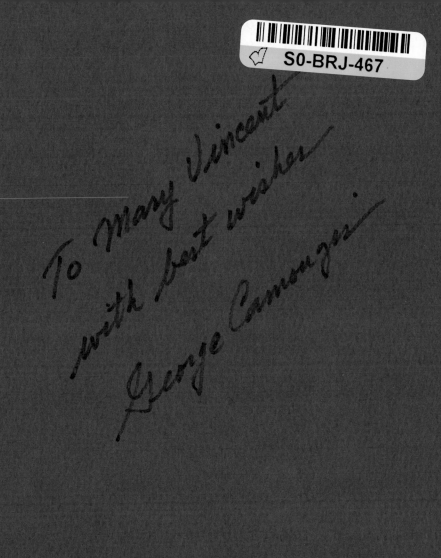

To Mary Vincent
with best wishes

George Camougis

Environmental Biology for Engineers

A GUIDE TO ENVIRONMENTAL ASSESSMENT

About the Author

George Camougis is currently president and research director of New England Research, Inc., a research and consulting firm established in 1968. He received his undergraduate degree from Tufts University and his M.A. and Ph.D. degrees in biology from Harvard University. He has over 25 years of experience in teaching, industrial research, and consulting. Dr. Camougis has published extensively and lectured throughout the world.

In his current capacity at New England Research, he serves as consultant to many private organizations and public agencies. Federal agencies for which he has consulted include the U.S. Army Corps of Engineers, the Department of Agriculture, the Department of the Interior, the Department of Transportation, and the Environmental Protection Agency.

Environmental Biology for Engineers

A GUIDE TO ENVIRONMENTAL ASSESSMENT

GEORGE CAMOUGIS

McGraw-Hill Book Company

New York   St. Louis   San Francisco

Auckland   Bogotá   Guatemala   Hamburg   Johannesburg
Lisbon   London   Madrid   Mexico   Montreal
New Delhi   Panama   Paris   San Juan   São Paulo
Singapore   Sydney   Tokyo   Toronto

Environmental Biology for Engineers: A Guide to Environmental Assessment

Printed in the United States of America
1234567890 KPKP 86543210

Library of Congress Cataloging in Publication Data

Camougis, George, 1930-
  Environmental biology for engineers.

  Bibliography: p.
  Includes index.
  1. Ecology. 2. Environmental impact analysis.
3. Environmental law. 4. Environmental engineering. I. Title.
QH541.C343   574.5'024363   80-17301

ISBN 0-07-009677-5

Contents

Preface

THIS BOOK IS INTENDED as a practical guide to environmental biology for engineers and other professionals concerned with environmental affairs and not trained primarily in the biological sciences.

Environmental biology concerns itself with the interplay among a broad spectrum of biological, chemical and physical phenomena. It encompasses both the aquatic and the terrestrial environments, and it includes the study of all groups of plants and animals. It is a field in which professional interaction with engineering specialties is increasing every year.

The past decade has become widely recognized as one of intensive efforts in the analysis and solution of environmental problems. Environmental issues have become focal points for the dynamic interactions among science, technology, law and public involvement in decision-making. Among the concerns and issues raised repeatedly by public agencies and private organizations are those that center on potential harm to living systems. These concerns have now become firmly established in environmental legislation. Much of the current effort in environmental biology is based on legislative mandates, and a recurrent theme in this book is the regulatory framework for these current efforts.

Three primary objectives are reflected in the format and organization of this book. First, the book presents a practical vocabulary in environmental biology. The text, the appendixes and the glossary all define terms used frequently in environmental studies. Next, the book discusses environmental biology in the context of both environmental engineering and environmental legislation. In that sense it relates the theory and practice of environmental biology to "real-world" applications in engineering projects. Finally, the book provides environmental engineers, project engineers, environmental planners and project managers with practical guidelines on the application of biology to environmental assessments.

This is not a book on how to write an environmental impact statement (EIS). Preparation of an EIS includes many disciplines and procedures not covered in this book. This book centers on the effective application of environmental biology to a wide range of engineering projects. Past failings in environmental projects have resulted mainly from failures in interdisciplinary efforts. Today there is still a critical need to improve the integration of environmental biology, environmental engineering and regulatory affairs. It is hoped that this book will contribute to the future refinements of such interdisciplinary efforts.

The substance of this book includes various contributions from several associates and colleagues, especially Edward J. Robbins, Paul A. Erickson, Norman H. Miner and Ronald D. Cheetham. Our experiences in team efforts on numerous projects have influenced the development of this book. The book has also been influenced by over 1500 engineers, scientists, planners and other professionals with whom we have come in direct contact during the past decade through our research, consulting and training activities at New England Research. These individuals represent government agencies, consulting engineering firms, industrial companies and utilities. Their objective skepticism, probing questions and sharp discussions have contributed enormously to the conceptual evolution of this book. Finally, it is a pleasure to acknowledge the contributions of Thaya A. McCrohon in the preparation of the book and Robert H. Dano for preparation of the graphics. Sincere thanks are offered to Sonia Sheldon of McGraw-Hill for editorial guidance. Any practical success that this book may enjoy will be due mainly to the valuable contributions from all these individuals.

George Camougis

Environmental Biology for Engineers

A GUIDE TO ENVIRONMENTAL ASSESSMENT

Chapter 1

Some Introductory Comments on Biology

Perhaps nowhere in modern society is the dynamic inter-play among the various disciplines so apparent as in the field of environmental engineering. The past decade, especially, has added much impetus to interdisciplinary ap-proaches to the analysis and solution of various problems. Yet despite the undisputed necessity for interdisciplinary efforts, anyone involved directly in such efforts recognizes immediately the enormous conceptual and communications bar-riers that exist among the various disciplines. The inter-faces between biology and engineering contribute their share of problems. This chapter is an attempt to describe the science of biology in a very broad overview, and to examine how some of the characteristics of biology are of direct relevance to environmental engineering.

THE SCIENCE OF BIOLOGY

Biology is the science of living things. It derives its origins from the word *bios,* meaning *life.* The prefix *bio-* is used commonly, and will be encountered frequently in this book (see Glossary).

All the living things on the earth are referred to by the collective term *biosphere.* The biosphere is global in magnitude, and is the largest unit of living mass (or *bio-mass).* As the focus changes to smaller systems, it becomes necessary to use other classifications for convenience in communication.

The first major subdivisions of biology are *botany,* the study of plants, and *zoology,* the study of animals. There are approximately 340,000 known species of plants in the world. The animals number well over 1,000,000 species

throughout the world. Thus biologists speak of the
plant kingdom and the *animal kingdom* to differentiate.

BIOLOGICAL NOMENCLATURE

   It becomes obvious that the enormous number of biological species dictates a system of biological nomenclature. This need for classification became apparent
long ago; for example, Aristotle developed a system of
biological nomenclature over 2000 years ago. The system
of classification as we know it today began with Linnaeus
(1707 - 1778), a Swedish scientist. The special field that
deals with the classification of living organisms is
called *taxonomy*. Additional details on biological
nomenclature may be found in Appendix 1.

COMPLEXITY OF BIOLOGICAL COMMUNITIES

   Another important factor of biology relates to the
complexity of biological communities. Most natural systems
are characterized by a wide variety of species that are
dynamically linked. It is possible to divide the biosphere
into progressively lower levels of complexity. This progressive reduction in complexity (and physical size) may
be illustrated as follows:

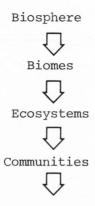

Biosphere

Biomes

Ecosystems

Communities

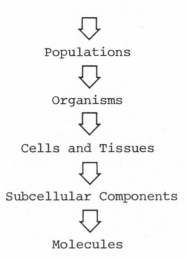

Populations

Organisms

Cells and Tissues

Subcellular Components

Molecules

These various levels of organization will be defined and discussed in subsequent sections of this book.

SPECIALIZED FIELDS OF BIOLOGY

As with most major disciplines, the science of biology encompasses many specialized fields. Indeed, it appears that more and more specialized subdisciplines are being formalized each year. One measure of the specialization within any science is the number of professional journals. Within the science of biology there are many thousands of journals now published on a regular basis; the total number is considerably more than in any other science. One practical consequence of this specialization is that biologists themselves have enormous communications problems. Most of the major specialties within biology are outlined and discussed briefly in Appendix 2.

ENVIRONMENTAL BIOLOGY

Environmental biology is a relatively new field. The central theme of environmental biology is the dynamic relationships between living organisms and the physical environment. There is considerable overlap among the various biological subdisciplines directly involved in environmental studies. A good example is seen in the fields of ecology, forestry and botany. Nevertheless there is now emerging a strong, separate identity for environmental biology. A wide spectrum of specialties are contributing vital, scientific roles to the field of environmental biology. It is not surprising, then, that today the environmental biology components of many projects are frequently studied by teams of biologists.

Environmental biology concerns itself with all levels of biological complexity, from cells to entire biomes. The more conventional approach in most engineering projects is to study the populations of dominant species within a specified area. The methods used and the degrees of quantitation proposed can vary considerably. However, great efforts are being made to standardize the methods used in field and laboratory investigations in environmental biology.

PRACTICAL RELEVANCE TO ENGINEERING

Many of these aspects of biology have direct, practical significance to the various efforts related to engineering projects. For example, biologists by intuition and training think of biological species in terms of their taxonomic (scientific) names. Moreover, they tend to be reductionist in their thinking, and therefore there is a strong tendency to classify to the lowest *taxon* (see Appendix 1). One of the stocks in trade of biologists is the use of taxonomic names. The degree of taxonomic detail has practical implications to budget, editorial format, regulatory requirements and the reactions of the lay public that must often read the technical reports on the project.

Another practical issue centers on the levels of complexity to be considered in a project. The easiest and most conventional approach is to make an inventory of the biological species present in a given area. However, the interpretation of the significance of these species often requires an assessment of dynamic interactions in populations and ecosystems. Such interactions cannot be studied as easily as the individual species. Consequently, there is a strong tendency to compile lists of birds, mammals, fishes, etc., and to ignore population dynamics, predator-prey relationships, effects of adverse meteorological cycles and other dynamic processes. Thus the levels of biological complexity that are considered can have practical significance to the overall approach, findings and conclusions that are reached in a given project.

The background and training of the biologists involved in a given project are also important. Biologists with undergraduate training tend to have general backgrounds, with some training in various specialties. Biologists with advanced degrees are almost always trained within a narrow specialty. For example, a biologist with an advanced degree and training in aquatic insects does not necessarily have any special background in desert plants. A broadly-trained individual without advanced training may often have more practical experience with a given group of organisms than someone with an advanced degree. On the other hand, a critical issue may require a specialist with considerable research experience in a specific field. Thus biology is no different from other disciplines, and the "universal biologist" does not exist. For this reason, the larger projects generally require the teams of biologists that we frequently see today.

Another factor of practical significance to engineers is the relatively non-quantitative nature of biology. There has been considerable progress in attempting to measure and quantitate certain biological phenomena. However, much of biology remains descriptive and non-quantitative, and this reality must be recognized and considered in the overall project (see Appendix 3).

In conclusion, many of the practical, day-to-day details of a project will depend greatly on the biologists that are working with the engineers. It is worthwhile to remember that the ultimate success of the project will depend upon how well the biological efforts are integrated with the engineering efforts and the regulatory requirements of the project.

Chapter 2

Environmental Legislation

Although research in various, specialized aspects of environmental biology has been conducted for decades, the environmental legislation in recent years has greatly intensified these research efforts both in the laboratory and in the field. Furthermore, most of these efforts are being applied to various engineering projects thought to have potential for environmental impact.

This chapter will attempt to provide an overview of some of the environmental legislation which provides the legal background for much of the current work in environmental biology. The assumption is made that the reader is already familiar with most of this legislation. Consequently, the primary focus of this chapter is to examine the relevance of this legislation to the biological components of environmental research and assessment efforts.

The United States government has often taken the lead in the passage and promotion of laws and regulations aimed at conserving or preserving the Nation's natural resources. In many cases the laws are applicable to Federally-owned lands and are therefore only examples for the states to follow. In other cases stronger action at the state level is forced through funding mechanisms or Federal review. Our concern in this chapter is with the role of Federal legislation in the protection of natural biological resources.

NATIONAL ENVIRONMENTAL POLICY ACT OF 1969 (NEPA)

On January 1, 1970 President Nixon signed the *National Environmental Policy Act of 1969* (Public Law 91-190; NEPA), and thus swung into motion one of the most powerful acts of

legislation in recent years. NEPA is the first piece of
Federal legislation that required coordination of Federal
projects and their impacts with all the national resources.
NEPA established a national policy for environmental affairs, mandated certain requirements for Federal government
decision-making process, and established the Council on Environmental Quality (CEQ) in the executive branch. The
Act is unique in its establishment of action-forcing procedures (Section 102) that must be followed prior to final
Federal decision-making. Regulations for implementation
have been promulgated by various agencies. These regulations have sought to integrate environmental considerations
into all levels of Federal decision-making. The effects of
a project on the biological resources is one key area that
the developing or permitting agency must address.

The regulations for the development of an Environmental
Impact Statement (EIS) generally require a full disclosure
of all the potential environmental consequences of a proposed action. Of particular concern to the biological area
is the emphasis on secondary as well as primary impacts.
It is not enough to recognize that there is a temporary increase in turbidity in a stream due to erosion at the
construction site. The effect of the turbidity on downstream food webs, fish populations, etc. must also be
considered. The concern for incremental degradation of
natural resources is expressed strongly in the regulations.
Congress recognized in the Act that impacts in the natural
resources area may not be subject to as rigorous a quantitative analysis as in engineering or economic considerations. Nevertheless, these considerations must be
integrated into the decision-making process. The courts
have taken notice of this requirement, along with the
Congressional statement of environmental objectives found
in Section 101. The courts have increasingly evaluated
cases brought under this Act in terms of the balance of
environmental considerations and the project's relationship
to the kind of environment Congress envisioned. It is
important to note that the various regulations are revised
periodically (e.g., CEQ, 1978).

PROTECTION OF BIOLOGICAL RESOURCES

Over the last twenty years, a series of Federal laws
have been enacted with the intent of protecting biological
resources and their habitat. Some of the older legislation
has been receiving increased enforcement attention from the
responsible agency. The passage of NEPA gave added rein-
forcement to the consideration of a proposed project's impact on protection afforded under this legislation.

The *Fish and Wildlife Coordination Act* was passed in
1958. Recent developments in Congress have stimulated the
U.S. Department of the Interior to increase its enforcement
efforts. The Act requires that a sponsoring or permitting
agency for a project that will divert, deepen, or alter a
stream or body of water must coordinate with the U.S. Fish
and Wildlife Service and the appropriate state agency. The
developing agency must agree with the wildlife agency on a
plan for the conservation and enhancement of the wildlife
resources affected. Increasingly, the U.S. Fish and Wildlife Service (USFWS) is concerned in this evaluation, not
only with what populations actually exist in the affected
area, but also with what kind of habitat exists and what
it could support.

The importance and need for protection of wetlands has
been recognized by the U.S. Congress in a series of acts.
Enabling legislation under the Migratory Bird Treaties with
Canada (1916) and Mexico (1936) requires the Department of
the Interior to acquire and protect wetlands along the flyways of migratory birds. *The National Wildlife Refuge System Administration Act of 1966* set up a central control for
the administration of the lands acquired as habitat. The
Water Bank Act (1970) enhanced the powers of the Secretary
of the Interior to preserve and improve wetlands as habitat
for wildlife. Finally, *Executive Order 11990* (May 24, 1977)
cited the objectives of NEPA and required that all agencies
must take an active role in minimizing the loss, destruction
or degradation of wetlands.

The U.S. Congress has also sought protection for plant
and animal species which have become threatened with extinction. *The Endangered Species Act of 1973* supplanted
all previous legislation on endangered species. It was
designed to provide a means whereby the ecosystems upon
which endangered species and threatened species depend may

be conserved, and to provide a program for the conservation of such endangered species and threatened species. The Act calls upon all Federal agencies to play an active role in this effort.

The USFWS has developed a three-step program for executing the Act's objectives:

1. Establish a list of threatened and endangered species (published in the *Federal Register*).
2. Determine and publish the geographical area which is the critical habitat.
3. Establish regulations and management policies for protection.

To comply with the Act, a developing or permitting agency may not approve any projects which will result in the destruction or adverse modification of critical habitats of listed species. The critical habitat may not be the entire habitat. It is that portion which is necessary for the maintenance and restoration of a species in the wild to a point where protection is no longer necessary. This requirement does not eliminate all projects in a critical habitat, only those projects or portions thereof that will adversely affect the functioning of the critical habitat. Discussions between the project sponsor and the USFWS will provide the ultimate requirements for ensuring that the proposed project is compatible with the legislation.

The list of threatened and endangered species of the animal kingdom includes mammals, birds, reptiles, fishes, snails, and clams. Protection for endangered plants was also included under the 1973 Act. Approximately 1700 plants were designated initially to receive protection under the Act. The Endangered Species Act is amended at periodic intervals. Revised lists of species are published in the *Federal Register*.

The U.S. Congress has also singled out particular species for protection, regardless of their population status. Examples of this legislation include the *Bald and Golden Eagle Protection Acts, Anadromous Fish Conservation Act, Wild Horses and Burros Protection Act*. Each of these requires the Department of the Interior to protect and enhance the species.

PROTECTION OF WATER AND MARINE RESOURCES

Protection of the aquatic habitat may be traced back as
far as the *Rivers and Harbors Act of 1899* (Refuse Act).
The Act was intended to preserve the integrity of navigable
waters, for the purpose of navigation. It was not until
1965 that the first piece of significant Federal legislation
was passed on industrial pollution of the waterways, the
Federal Water Pollution Control Act of 1965. This Act concentrated on interstate waters (which are only 14 percent of
all U.S. waters), and used receiving water quality standards
as a means of control. By the late 1960s, it became evident
that this approach was too cumbersome to accomplish the objectives. For a while the Federal government tried to use
the *Refuse Act* to enforce pollution control. The *Federal
Water Pollution Control Act Amendments of 1972* grew out of
the failure of these efforts.

In passing the Act, Congress stated that it wished to
"restore and maintain the chemical, physical and biological
integrity of the Nation's waters". Congress set national
goals including the following:

1.   The achievement of a water quality level which
     "provides for the protection and propagation
     of fish, shellfish, and wildlife" and for
     "recreation in and on the waters" by July 1, 1983.
2.   The elimination of the discharge of pollutants into
     U.S. waters by 1985.

The strategy for accomplishing these goals shifted to control of point source discharges at the end-of-the-pipe with
minimum standards. However, other sections of the Act required the development of water quality goals based on end-
use and plans for their achievement for the receiving
waters. These end-uses relate to human and wildlife needs.
New projects must fit these goals and plans.

The Congress has extended its concern for the protection of biological resources in the ocean through *The Marine
Protection, Research and Sanctuaries Act of 1972*. This Act
regulated ocean dumping, provided funds for research to end
all ocean dumping, and created a system by which marine
areas as far out as the edge of the continental shelf and
the Great Lakes could be designated as sanctuaries and
protected from incompatible activities. The U.S. Army Corps
of Engineers and the U.S. Environmental Protection Agency
(EPA) have responsibility for assessing the impacts and

controlling a proposed dumping project. The *Deepwater Port Act of 1974* extended the Congressional concern that marine projects must be compatible with the protection of both marine and shore biological resources.

LAND USE PLANNING AND CONTROL

The environmental concern of the 1970s has brought widespread attention to the importance of land use. The control of land use has been almost exclusively under the jurisdiction of local government through zoning laws. The normal concerns were compatibility of use to maintain the economic value of the land. Since the late 1950s the Federal and state governments have begun to impose on this area. The decisions at all regulating levels and in the courts recognize more than an economic interest. Public environmental and aesthetic interests are often juxtaposed with the private right to develop property.

At the Federal level, there are a number of regulations or acts that have brought the Federal government into land use planning. The *Department of Transportation Act* required the preservation of parkland and refuge areas in highway planning. NEPA brought environmental factors into the permitting process. The *Clean Air Act* (1970) as it is now being interpreted under the "no significant deterioration" doctrine has placed constraints on industrial development. Section 208 of the *Federal Water Pollution Control Act Amendments of 1972* provides for area-wide waste treatment management plans. The control of non-point source pollution under these plans will depend on land use planning.

The Federal government has also taken direct action. The *Coastal Zone Management Act of 1972* provided Federal funds to assist states to develop land use plans in the coastal area. Congress recognized the coastal zone as "ecologically fragile and consequently extremely vulnerable to destruction by man's alterations" and an "urgent need to protect and to give high priority to natural systems in the coastal zone". The Act applies to the states bordering on the Great Lakes as well as those on the ocean. Some 30 states are eligible for this assistance.

TOXIC AND HAZARDOUS SUBSTANCES

Toxic substances in the environment are becoming an increasingly important concern to the general public. The suspension of DDT, the concentration of PCBs (polychlorinated biphenyls) in fish in the Hudson River, and the Kepone incident in the James River have all focused public attention on the problem. A major difficulty in the control has been the definition of hazardous and toxic substances. Early toxic substance legislation centered on foods *(Food, Drug, and Cosmetic Act)* and pesticides *(Federal Insecticide, Fungicide and Rodenticide Act)*. The passage of the *National Environmental Policy Act of 1969* provided a broad conceptual base for assessing the interrelationship between humans and their environment. The release of toxic substances into the environment is an issue to be addressed as part of the mandate under NEPA. Many other Federal environmental laws have addressed the issue in a narrower context. Agency regulations which interpret these often have an effect upon the construction, operation and maintenance activities of a proposed project.

The *Federal Water Pollution Control Act Amendments of 1972* seek to control toxic pollutants through several areas. Under Section 208 an overall area-wide planning process is created to handle toxic substances in the waste stream. The plans developed for each local area may govern the handling of wastes from a project in both its construction and operational phases. Section 307 of the Act addresses the subject of toxic pollutants in point source effluent directly. The EPA must publish a list of substances and effluent standards for these. Dredged material is specifically included in the list of potential sources of toxic pollutants. The EPA is currently studying the ecological and health effects of 65 potentially toxic pollutants. These studies will become important input for the evaluation of the probable effects of toxic materials that may be associated with the effluent of a proposed project. Section 311 of the Act requires the definition of hazardous substances which, when discharged into the waters of the U.S., will present an "imminent and substantial danger to public health or welfare, including, but not limited to, fish, shellfish, wildlife, shorelines and beaches". Some 300 substances, including oil, have been designated for regulation. Under Sections 403 and 404 of the Act, and under

Section 102 of the *Marine Protection, Research and Sanctuaries Act of 1972,* the Corps of Engineers and the EPA control the dumping of dredge materials and other wastes into the navigable waters and the ocean. The toxicity of the wastes is a key concern in the criteria that were developed for the evaluation of proposed permits.

The *Safe Drinking Water Act of 1974* expressed Congressional concern for toxic substances in drinking water supplies. Under the requirements, the U.S. EPA must issue regulations on drinking water standards and on minimum standards for state programs for the control of underground injections of toxic or hazardous materials. The potential effects of toxic substances associated with a proposed project on a watershed, a surface drinking water supply, or the ground water are major issues in current projects.

The *Clean Air Act Amendments of 1970* also contained authority for the EPA to set emission standards for hazardous air pollutants. The list includes asbestos, beryllium, mercury and vinyl chloride. In setting these standards, the EPA is concerned only with health considerations. Clean-up technology and cost are not valid considerations in the setting of the standards.

The *Toxic Substances Control Act* was passed in 1976. This important Act is directed towards the evaluation and control of manufactured substances which are potentially toxic. The EPA has prime enforcement responsibility for the Act. However, much of the regulatory authority for toxic substances remains fragmented. Several Federal agencies must be consulted to ensure that a proposed project meets all requirements.

CONSERVATION, PRESERVATION AND RECREATION

The earliest legislation for protection of natural and recreational areas came in 1872 when the world's first National Park, Yellowstone, was established. The National Park Service was established in 1916 to administer the growing system of parks.

In passing the *Wilderness Act* in 1964, Congress recognized the value of preserving certain areas that would be left "unimpaired for future use and enjoyment as wilderness". These areas are to be "undeveloped Federal land retaining its primeval character and influence without permanent improvements or human habitation". The Wilderness

Preservation System was established to administer the program. Over 30 areas have been designated in the system, while an equal number are under consideration for inclusion in the system. Further evidence of the Congressional concern for preservation of natural resources is found in the *Wild and Scenic Rivers Act* (1968). This Act authorized the Departments of the Interior and Agriculture to study certain rivers and to acquire land along them for protection from development. Management plans for the river are based on the type of recreational opportunity to be provided. *The National Trails System Act* (1968) provided for the development and protection of hiking trails. Congress set up a system of national scenic trails and national recreational trails. The latter is intended to be close to urban areas for maximum public access.

The U.S. Congress also encouraged the states to plan and develop outdoor recreational opportunities by providing matching funds under the *Land and Water Conservation Fund Act of 1965*. Once the plans have been set and land areas have been purchased under the Act, it requires the compliance of the Secretary of the Interior to alter the land use to make way for a project. The Secretary will only agree to the change if land of equal value that is consistent with the state's outdoor recreation plan is provided.

SUMMARY

The broad scope of Federal legislative action concerned with the protection of natural and biological resources is evident in Table 1. Project development which requires the acquisition of permits from a Federal agency should only be undertaken with a full knowledge of how the project may be affected by these laws. Moreover, general alertness to frequent revisions of guidelines and regulations is necessary. Coordination with the appropriate agencies may be the least requirement. The design and construction of mitigation and/or enhancement features may be a more likely requirement as the project increases in size and scope. This legislation is the foundation upon which appropriate biological studies must be built early in the project planning cycle.

TABLE 1.   *Summary of Federal Acts Associated with Natural Resource Protection.*

GENERAL
    Environmental Quality Improvement Act of 1970
    National Environmental Policy Act of 1969

PROTECTION OF BIOLOGICAL RESOURCES
    Anadromous Fish Conservation Act
    Bald Eagle Protection Act
    Commercial Fisheries Research and Development
        Act of 1974
    Endangered Species Act of 1973
    Federal Aid in Fish Restoration Act
    Federal Aid in Wildlife Restoration Act
    Fish and Wildlife Act of 1956
    Fish and Wildlife Coordination Act of 1958
    Golden Eagle Protection Act
    Marine Mammal Protection Act
    Migratory Bird Conservation Act
    National Wildlife Refuge System Administration
        Act of 1966
    Wild Horses and Burros Protection Act

WATER RESOURCES PROTECTION AND DEVELOPMENT
    Federal Power Act
    Federal Water Pollution Control Act Amendments
        of 1972
    Federal Water Project Recreation Act
    Rivers and Harbors Act of 1899 (Refuse Act)
    Safe Drinking Water Act of 1974
    Water Bank Act
    Water Resources Research Act of 1964
    Watershed Protection and Flood Prevention Act

MARINE RESOURCES
    Deepwater Port Act of 1974
    Estuarine Areas Study Act
    Marine Protection, Research and Sanctuaries Act of 1972
    Ports and Waterways Safety Act of 1972

TABLE 1.   *(Continued)*

LAND USE PLANNING AND CONTROL
    Clean Air Act Amendments of 1970
    Coastal Zone Management Act of 1972
    Department of Transportation Act
    Federal Water Pollution Control Act Amendments of 1972

TOXIC AND HAZARDOUS SUBSTANCES
    Clean Air Act Amendments of 1970
    Federal Environmental Pesticide Control Act of 1972
    Federal Water Pollution Control Act Amendments of 1972
    Marine Protection, Research and Sanctuaries Act of 1972
    Pesticides Research Act
    Resource Conservation and Recovery Act of 1976
    Rivers and Harbors Act of 1899
    Safe Drinking Water Act of 1974
    Toxic Substances Control Act

CONSERVATION, PRESERVATION AND RECREATION
    Federal Water Project Recreation Act
    Land and Water Conservation Fund Act of 1965
    National Trails System Act
    Water Bank Act
    Wild and Scenic Rivers Act
    Wilderness Act of 1964

Chapter 3

Major Biomes

Various regions of the earth are dominated by specific
types of vegetation that are rather characteristic of that
area. Physical and geographic factors normally play the
major role in the determination of which plant species will
grow in the area. Extensive areas characterized by typical
plant communities may cover thousands of square miles.
These areas are often referred to as *biomes* (see page 2).
The plant community provides the dominant biological feature in a biome. Biologists recognize various types of
biomes, although there is often lack of precise agreement
in the specific terminology. The major biomes of the
North American continent are outlined in this chapter.
Biomes are very large communities, and include both plants
and animals (see Figure 1).

TUNDRA

Stretching along the northern regions of the North
American continent from Alaska to eastern Canada, there is
a vast area of mat-like vegetation referred to as the
arctic tundra. The climate is characterized by low
average temperatures, short growing seasons, and low
annual precipitation. Dominant vegetation of the tundra
includes lichens, mosses and sedges. The tundra is noted
for its lack of tall vegetation such as trees. Growth of
vegetation is slow compared with other biomes. The tundra
biome is rather fragile, and consequently it receives much
attention in the environmental studies. Concern for the
tundra was one of the major issues in the Alaska oil pipeline project.

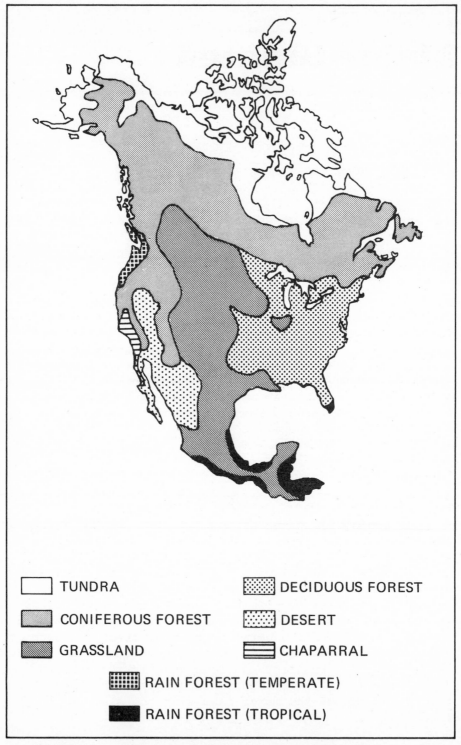

FIGURE 1.   Diagram of General Geographic Ranges of
Major Biomes in North America.

At lower latitudes, and at higher elevations, various mountains may have a similar vegetation. This is referred to as the *alpine tundra* or *alpine meadow*. In contrast with the arctic tundra, the alpine tundra may grow in areas with more annual precipitation and longer growing periods. Also, the alpine tundra does not normally have the permanently frozen deeper layers called the *permafrost* which are characteristic of the arctic tundra. The alpine tundra is found extensively throughout North America.

The tundra is a harsh environment for animals. Nevertheless, a number of larger species exist successfully in this biome, including the caribou, the arctic fox, the arctic hare, and the lemming. Other animals, including birds and insects, can also survive in the tundra biome.

CONIFEROUS FOREST

Generally south of the arctic tundra, and just as extensive from coast to coast, there is a very large biome in which coniferous (evergreen) trees make up the dominant vegetation. This area is called the *northern coniferous forest* or the *taiga*. Various species of pine, hemlock, and cedar are the dominant vegetation. In addition to this *overstory* vegetation made up of evergreen trees, there is an *understory* made up of shrubs and other plants. Coniferous forests are generally characterized by longer growing periods and more precipitation than the tundra.

Other coniferous forests may be found in more temperate climates. Also they may be found on mountain slopes. Such forests will have different species of evergreen trees that make up the dominant vegetation. In the somewhat more southern regions, the coniferous forest may be intermixed with hardwoods to form a mixed *coniferous-deciduous forest*.

Many species of animals live in coniferous forests, including some of the more commonly known wild animals. These include such species as bears, wolves, elks, moose, otters, lynx, hares, and squirrels. Coniferous forests also support many birds and invertebrate animals.

DECIDUOUS FOREST

Deciduous trees, in contrast with coniferous trees, shed their leaves on a seasonal basis. Deciduous trees generally require a longer growing period, and tend to be in areas south of the coniferous forest. There are many species of deciduous trees in North America, and it is possible to distinguish many associations of dominant species. The major regions of the *temperate deciduous forest* include the midwest and the east. The major vegetation of this biome includes various overstory species of beech, maple, oak, and hickory. In addition, there is usually a lush understory of shrubs and herbs. Along the western regions of the continent, the deciduous trees are less numerous. They are often interspersed with other types of vegetation. Thus the temperate deciduous forest biome is generally characteristic of the eastern half of the temperate region of North America. There are frequent overlaps with the coniferous forest. Sometimes the term *ecotone* is used to define the region where there is a transition from one type of community to another. Thus the deciduous forest, the coniferous forest, and the tundra are separated by ecotones or transitional areas.

Animals of the temperate deciduous forest include deer, bear, squirrels, foxes, rabbits, and bobcats. Some animals of the coniferous forest will also extend to the deciduous forest, especially in the ecotone between the two biomes. Smaller animals are common among the understory plants. Numerous birds are found in the temperate deciduous forest.

GRASSLAND

Another extensive biome in the North American continent is the *grassland* biome. This biome makes up the huge continental basin generally east of the Rocky Mountains and west of the deciduous forest. It extends from central Canada into sections of Mexico. General climatic features of the grassland biome include hot summers, cold winters, relatively little annual precipitation, and generally windy conditions. The grassland biome of North America is also called the *prairie*. The terrain of the grassland biome is generally flat or gentle slopes. The dominant vegetation consists of various species of grasses. Specific zones with

different species may be present. One characteristic of
the grasses is extremely deep penetration of the root system
into the soil (several feet). Other important plants include legumes. Grasses may grow several feet high, depending upon species, and thus dominate the vegetation of
the grassland biome.

Among the animals associated with the grasslands are
those adapted to grazing. These include bison, pronghorn
antelope, and deer. Other species include animals that bur-
row, such as ground squirrels and gophers. Coyotes and
hawks are important predators.

CHAPARRAL

In areas with mild climate and relatively wet winters
but dry summers there exists a community known as the
chaparral biome. The vegetation of the chaparral consists
of hardy trees and shrubs with thick, evergreen leaves.
The chaparral biome is not extensive and is discontinuous.
The larger regions are found in California, Arizona, New
Mexico, and Mexico.

Animals in the chaparral include rabbits, rodents, lizards, and snakes. In some sections there may be larger
mammals such as the mule deer. Small birds and birds of
prey are also present.

DESERT

Closely related to the chaparral is the desert biome.
Here there is even less annual precipitation. The temperature is hot during the day and cold during the night. Evaporation is high. Plants are adapted to the arid conditions,
and include cacti and yuccas. However, shrubs and various
flowers may also be present, especially in relatively moist
areas. During brief periods of rainfall, ephemeral plants
may add brilliance and variety to the desert plant life.
The desert biome is located in the southwestern part of the
United States and in parts of Mexico.

Animals in the desert biome are similar to those of the chaparral. Various small mammals such as rodents are present, but these are especially adapted for the harsh environment. Burrowing is one form of adaptation to desert conditions. Other animals include snakes, lizards, and various insects.

OTHER BIOMES

Other biomes of lesser geographic extent also exist in North America, especially in coastal areas. Localized areas with high rainfall support *rain forest* biomes. Southern Florida is characterized by a *tropical rain forest*. Sections of the Pacific Coast have a *temperate rain forest*. Rain forests have very high annual precipitation and high humidity. Dominant species in rain forest vary greatly, and are determined by geography, temperature, and other features.

On a global basis, similar biomes exist in other continental land masses. The tundra stretches for thousands of miles from northern Europe into Siberia. The northern coniferous forest is also found in Europe and Asia. The temperate deciduous forest dominates most of central Europe. Tropical deciduous forests exist in South America, Asia and Australia. Grasslands are extensive in Asia, Africa and Australia. Extensive deserts are also found in Asia, Africa and Australia. A chaparral biome borders much of the Mediterranean basin. Plant and animal species within these biomes will differ when compared to North America. However, the basic organizations and community structures are similar.

Some authors include the sea as a *marine* biome, although the concept of dominant vegetation is, strictly speaking, more applicable to the terrestrial environment. In the marine biome plants do not exert the controlling influence as in the case of the terrestrial biomes. The physical and chemical properties of the marine environment exert the major influence on living organisms. The open water is a continuous, rather stable environment. Along the coastal zones, however, conditions may vary considerably.

MAJOR DETERMINANTS

It becomes clear from the foregoing discussions that
the biomes vary greatly. One way of viewing these various
biomes is to see them as the integration of various factors.
Among the major determinants of the characteristic vegetation of a biome are the following:

latitude	solar radiation
altitude	geologic conditions
annual precipitation	wind conditions

Thus the various biomes tend to have biogeographic distributions which reflect the major physical conditions within
the continental land mass. The major plant species in turn
determine all the other biota within the biome. It is important to remember that the species of plants and animals
may be quite numerous. For example, the deciduous forest
biome may have several hundred species of animals, even in
relatively localized areas. The examples given in this
chapter are only intended to present major species, and not
a typical inventory for each biome.

Chapter 4

The Ecosystem Concept

Ecology is the specialized field within biology concerned with the interactions among biotic and abiotic elements. These elements combine to form a system referred to as an *ecosystem*. Ecosystems are generally complex, and are characterized by the flow of matter and energy, as well as by biological populations within the defined area of the system. The geographic limits of any specific ecosystem are difficult to define. Consequently, the ecosystem provides as much a conceptual approach as it does a spatial approach to the study of biological populations within a given area. The ecosystem approach provides a powerful, conceptual tool which can be applied to environmental studies.

FOOD CHAINS

All ecosystems have well-defined components, and include organisms specialized to perform certain functions. These specialized functions within ecosystems are called *niches*. Certain organisms can capture solar energy, take in inorganic *nutrients* (e.g. phosphorus, nitrogen, carbon) from the environment, and through the process of *photosynthesis* produce organic molecules (e.g., sugars). Such organisms are called *producers,* or sometimes *primary producers*. Green plants are primary producers. Other organisms can consume organic materials and use the chemical energy in such materials to maintain their own bodies and grow. Such organisms are called *consumers*. Animals are consumers.

There are two types of consumers: (1) *herbivores* which feed directly on producers (i.e., green plants), and (2) *carnivores* which feed on other consumers (i.e., animals). Herbivores are also called *primary consumers*. Animals that feed on primary consumers are also called *secondary consumers* or *first order carnivores*. First order carnivores are in turn consumed by *second order carnivores;* and so the sequence may continue with several *higher order carnivores*. It becomes obvious that this is a system of linking animals (consumers) in *predator-prey* relationships, with the predator eating a lower order consumer as its prey. Such a system is called a *food chain*. This sequence may be illustrated as follows:

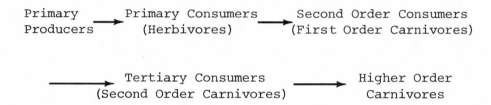

Primary Producers → Primary Consumers (Herbivores) → Second Order Consumers (First Order Carnivores) → Tertiary Consumers (Second Order Carnivores) → Higher Order Carnivores

Because the primary consumers are often called *grazers*, the sequence illustrated above is often called a *grazing food chain*.

Two other terms are frequently used in the terminology of ecosystems. *Autotrophs* (self-nourishing) refers to organisms, such as green plants, which can synthesize organic materials from inorganic nutrients. Primary producers are autotrophic organisms. *Heterotrophs* (other-nourishing) refers to organisms which ingest other organisms or parts of organisms in order to acquire the complex molecules they require for maintenance and growth. All consumers are heterotrophic organisms.

FOOD WEBS

Most ecosystems do not operate exclusively with the simple food chains outlined above. Consumers such as first and second order carnivores generally have multiple species that they prey upon. When one considers that there are often many consumers (hundreds of species) in an ecosystem, it becomes apparent that many predator-prey relationships exist. Viewed from another perspective, many alternate pathways exist. This concept can be illustrated in the simple diagram below:

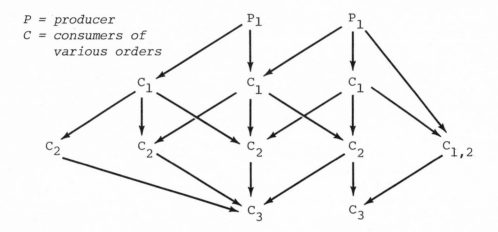

P = producer
C = consumers of
    various orders

Instead of a single food chain, there now exists a series of interconnecting chains. Of interest is the organism on the right of the diagram designated as $C_{1,2}$. This organism can eat both producers (P_1) and primary consumers (C_1). Thus it can be looked upon as an *omnivore* (mixed feeder). Such a complex system is referred to as a *food web*. Most systems have such food webs rather than simple food chains.

THE ROLE OF DETRITUS

Not all organisms, of course, are eaten by other organisms in a predatory sequence. Many organisms die from various causes including naturally-induced environmental stress. These organisms decay and are broken up by various chemical and mechanical forces. Also, all organic material ingested by consumers is not assimilated into the body chemistry of the consumers. Much of it is excreted to form part of an organic pool. All this organic matter is called *organic detritus* (or detritus). Even though it is not alive, detritus represents a rich source of organic material. Some consumers can ingest this detritus and derive the organic nourishment they require. Such organisms are called *detritivores*. Detritus-feeders, in turn, are eaten by carnivores,

and the usual food chain is formed. This is called a
detritus food chain. Some of the detritus, however, is
acted upon by *decomposers*. Decomposers are simple organisms (bacteria, fungi) which break down detritus into
inorganic nutrients (e.g., nitrates, phosphates, carbon
dioxide), at the same time deriving energy for their own
needs. The role of detritus in the ecosystem is illustrated
in Figure 2.

THE ROLE OF ENERGY

Ecosystems are also characterized by the flow of *energy*.
The ultimate source of energy for all living systems is the
sun. Solar energy is absorbed by primary producers and
converted into chemical energy. As organisms consume prey
in the food web, the chemical energy represented by the
bodies of the prey is utilized. However, only about 10% of
the energy is chemically incorporated into the new body
biomass of the predator; the remainder (90%) is eventually
lost. Organic detritus eventually will be broken down, and
here again, most of the energy is released. This energy is
eventually dispersed as unavailable heat energy. The energy
flow in an ecosystem is always in one direction. This is
a fundamental principle of thermodynamics, and ecosystems
obey the laws of thermodynamics. The flow of energy in the
ecosystem is illustrated in Figure 2.

TROPHIC RELATIONSHIPS

It becomes clear, now, that an ecosystem is highly complex, involving many organisms, the flow of energy, and the
accumulation of detritus. One way to view the ecosystem is
to see it as a hierarchy of trophic levels. This is illustrated in Figure 3 as a *pyramid of numbers*. From our
previous discussions it is reasonable to predict that the
number of individuals at each trophic level will become
smaller within any system. Since most of the energy is
lost, the prey populations required to maintain the predator
populations of the next higher trophic level must be greater
than that of predator populations. Thus a pyramid of
numbers exists. There is, however, another interesting
phenomenon in trophic relations. As one proceeds upward on

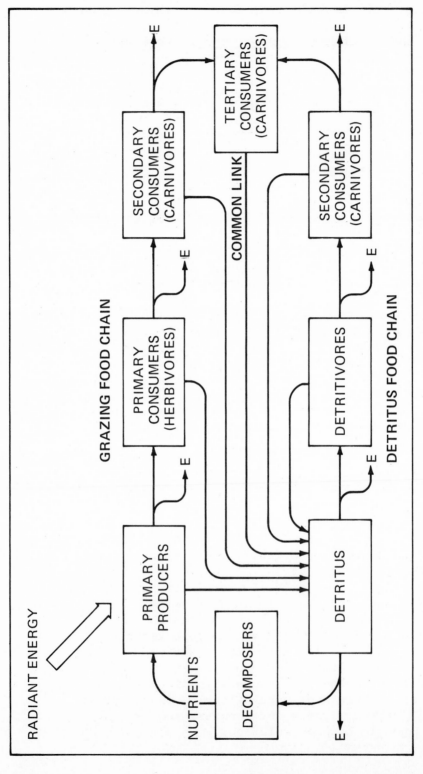

FIGURE 2. *Diagram of an Ecosystem, Showing a Grazing Food Chain, a Detritus Food Chain and the Flow and Loss (E) of Energy.*

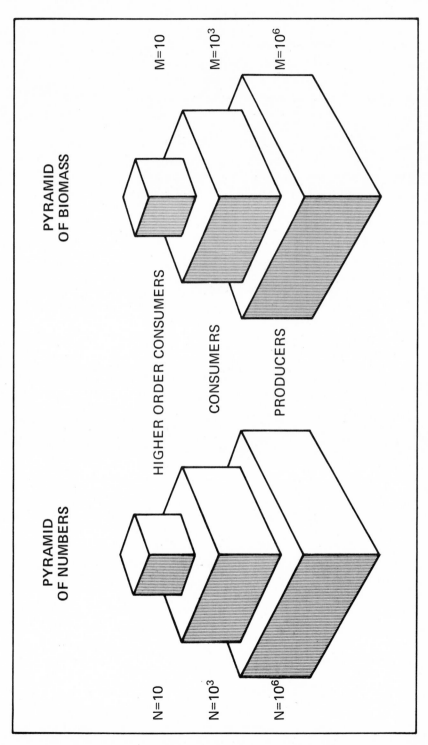

FIGURE 3.   *Diagrams Illustrating the Pyramid of Numbers (N) and the Pyramid of Biomass (M) with Various Trophic Levels.*

the pyramid, the organisms in the successively higher
trophic levels tend to get larger. The higher order carnivores are generally large. Therefore, a pyramid of numbers
does not always reflect the true relationships among the
various trophic levels. For that reason, ecologists also
use the concept of the *pyramid of biomass*. The pyramid of
biomass parallels the pyramid of numbers, in that the biomass of progressively higher trophic levels also tends to
decrease. This is illustrated in Figure 3. The ecologists
have another term that is used frequently in this context.
The amount of living material or biomass in a given trophic
level (or population) is called the *standing crop*.. This
term will be used again in later sections of the book.

THE ECOSYSTEM

     It is now convenient to summarize the various components
of an ecosystem. Four major components are present, and include the (1) abiotic elements (nutrients, water, air), (2)
the producers (or autotrophs), (3) the various consumers (or
heterotrophs), and (4) the decomposers. These are illustrated in Figure 2. Each organism plays a specific role,
or, to use another term, occupies a *niche* (role) within
the system. Note that the system illustrated contains both
grazing and detritus food chains. Note also the flow of
energy. Although energy flow is uni-directional and cannot
be retrieved, matter in ecosystems can be cycled over and
over again. Thus a carbon atom can become part of a plant,
part of an animal (consumer), part of the organic detritus
pool, degraded into inorganic carbon (CO_2), and become incorporated again into plant biomass. Consequently, we
speak of the constant cycling of matter (e.g., nutrients)
into *biogeochemical* cycles. And, since all ecosystems are
ultimately coupled, the entire biosphere (Chapter 1) can
be looked upon as part of the huge biogeochemical cycles of
the entire earth.

Chapter 5

The Terrestrial Environment

In this chapter the focus is on important biological concepts applied to the terrestrial environment. One important point to remember is that terrestrial systems, like all ecosystems, follow the same basic structure and dynamics discussed in Chapter 4. Consequently, we will expect the same basic components discussed above.

Also in this chapter we will attempt to progress from the more theoretical aspects of terrestrial ecosystems to some topics frequently of interest in engineering projects. Thus the discussions will include habitat, terrestrial wildlife, and the potential impacts of man.

TERRESTRIAL ECOSYSTEMS

We can now apply the same basic components of any ecosystem to the terrestrial environment. The abiotic elements contribute nutrients (phosphorous, nitrogen, carbon) and water which are combined with solar energy in primary producers (trees, shrubs, forbs, grasses) to synthesize the necessary organic components. As we saw in our discussions of biomes (Chapter 3), the dominant vegetation (producers) will be determined by numerous factors (e.g., climate). The producers will in turn influence the types of consumers that are present. Much of the organic material (especially dead plants, leaves) eventually falls to the ground as *litter* (detritus). Various consumers (herbivores, carnivores, omnivores) are present in all terrestrial systems. Finally, the decomposers are constantly at work breaking down the litter into nutrients that leach back into the soil. These nutrients are absorbed by the roots

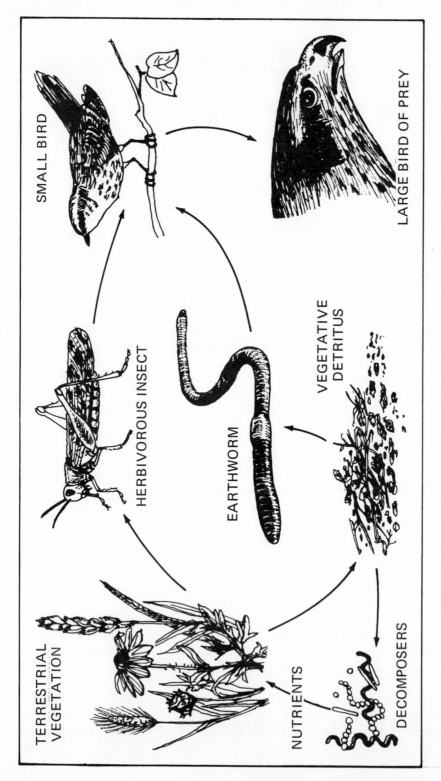

SMALL BIRD

LARGE BIRD OF PREY

HERBIVOROUS INSECT

EARTHWORM

VEGETATIVE DETRITUS

TERRESTRIAL VEGETATION

NUTRIENTS

DECOMPOSERS

FIGURE 4.   *Diagram of a Simple Terrestrial Ecosystem.*

of producers and new plant biomass is produced. Thus the
biogeochemical cycles operate constantly. Figure 4 is a
diagram of a terrestrial ecosystem in a geographic region
with a temperate climate.

The geographic location is not important for the
ecosystem concept and its application to the terrestrial
environment. Geography is important, however, when it
comes to specific lists of *flora* (plants) and *fauna* (animals). Such lists are important for environmental studies,
and since most engineering projects are site-specific,
lists of *biota* (plants and animals) are frequently required for such projects. These requirements will be discussed in more detail in subsequent sections.

In summary, we can outline the components of terrestrial
ecosystems as follows:

Components	*Examples*
Abiotic elements	Surficial geology (including soil types) Hydrology Meteorology
Producers (terrestrial flora)	Trees Shrubs Forbs (includes flowers) Grasses
Consumers (terrestrial fauna)	Invertebrates (e.g., snails, insects) Amphibians Reptiles Birds Mammals
Decomposers	Fungi Bacteria

ECOLOGICAL SUCCESSION

A very important concept in terrestrial ecology is that
of *ecological succession* of plants and animals in a specific
region. Most of the area within a biome is composed of
vegetation and biotic communities in a relatively stable
condition of long standing. However, such a stable

condition did not always exist. The temperate deciduous
forest is a good example to illustrate this concept. The
forest ecosystem is preceded by a series of steps in succession before the final, stable community is reached. Let
us begin by visualizing our area denuded of vegetation.
This area provides excellent opportunity for the spores and
seeds of vegetation to grow. First come grass and weeds.
When these are well established, the soil is receptive to
the growth of shrubs. Eventually shrubs give way to the
seedlings of trees. Initially the pine community is established. This is followed by deciduous trees such as
oak and eventually by the beech-maple forest. The animals
associated with each stage will also change. The entire
process is called ecological succession, and the final
stage is called the *climax stage*. The vegetation of the
climax stage is called the *climax vegetation,* and the entire
plant and animal community is called the *climax community*.

Complex physical and chemical changes are associated
with such successions. The accumulation of soil, humus
and litter occurs. The ability to retain moisture is enhanced. The penetration of sunlight is changed with successive changes in overstory and shrub vegetation. The
action of wind is altered. Surface hydrology is also altered. As these abiotic features change, the biota change.
Thus the process of ecological succession creates conditions which lead to successive changes until the final
stage (climax) community is established.

SOME SPATIAL CONSIDERATIONS

There are several concepts with spatial or geographic
considerations which are also important in ecology. Most
animals are distributed geographically within a biome (or
possibly biomes) where conditions permit them to compete
successfully. The large area within which a given species
is found is called its *geographic range*. The much smaller
area within which a single animal is found is called the
home range. For example, the geographic range of the white-
tailed deer is many thousands of square miles, while the
home range of a single deer may be on the order of square
miles. Smaller animals have correspondingly smaller home
ranges.

Many vertebrate animals (e.g., birds) will actively defend a specific area or *territory* against intruders, including members of the same species. This phenomenon is often called *territoriality* or *territorial behavior*. The significance of defending the territory is related to food requirements, and is usually associated with nesting and rearing of the young. The territory is usually smaller than the home range. Territorial behavior is common among vertebrates, including aquatic species such as turtles and fish.

HABITAT

Another important concept with spatial implications is that of *habitat,* or the physical location where biota are found. As we have seen, ecosystems are characterized conceptually by the flow of energy and matter and by complex interrelationships among biota. Often it is difficult to delineate the boundaries of an ecosystem. Therefore, ecosystems are not easily visualized by the public. However, the layperson is very likely to understand and have a feeling for a pine forest, an open field, and a river valley. Also professionals involved in the management of natural resources tend to think in terms of specific habitats. In the analysis of a region, therefore, it is useful to discuss types of habitat. Habitat assessment can be a complex, technical task. However, it is useful to classify the general habitats near a project area by a simple system such as:

> *forested areas* (deciduous, coniferous, mixed)
> *open fields* (commonly abandoned farms)
> *agricultural land*
> *semi-urbanized areas* (limited vegetation)

Note that this is similar to an analysis of *land use*. In fact, such a simple analysis of land use can tell us a lot about the habitat that is available for biota. The vegetation of the region provides both food and shelter for the wildlife in the area.

TERRESTRIAL WILDLIFE

Also of practical value in environmental studies is the
analysis of *wildlife*. The term wildlife, technically
speaking, includes all animals (aquatic and terrestrial),
although commonly it is used in the context of terrestrial
mammals. For our purposes we will make the distinction,
and apply the term terrestrial wildlife to this discussion.

Local interest in terrestrial wildlife is often expressed
by such diverse groups as naturalists, field sports groups,
bird clubs, and schools. In addition, departments of fish
and wildlife (state and Federal) frequently focus their
interest on local wildlife. Therefore, there is a need to
study these animals. This is frequently done through the
compilation of an inventory of certain classes of vertebrates (see Appendix 1), including:

amphibians   birds
reptiles   mammals

Invertebrate animals are commonly ignored in most terrestrial wildlife surveys. The same is true with decomposers. The reasons for these common omissions are complex,
but certainly one reason is that the general public and the
wildlife agencies have historic interests in the larger
animals (especially mammals). Nevertheless, the ecosystem
concept tells us that terrestrial ecosystems could not
function as integrated, dynamic systems without all the
components being present. Nor is it possible to make
reasonable cause-and-effect predictions of how a system will
respond to a given project without the ecosystem overview.

THREATENED AND ENDANGERED SPECIES

The legislative context for the importance of threatened
and endangered species was outlined in Chapter 2. Original-
ly the term "rare and endangered species" was used; however, since the *Endangered Species Act of 1973* (P.L. 93-
205), the terminology has been formalized as follows:

1.   *Species* - The term "species" includes any sub-
     species of fish or wildlife or plants and any
     other group of fish or wildlife of the same
     species or smaller taxa in common spatial arrange-
     ment that inter-breed when mature.
2.   *Threatened* - The term "threatened species"
     means any species which is likely to become
     an endangered species within the foreseeable
     future throughout all or a significant portion
     of its range.
3.   *Endangered* - The term "endangered species"
     means any species which is in danger of ex-
     tinction throughout all or a significant
     portion of its range other than the species
     of the Class Insecta determined by the
     Secretary to constitute a pest whose protec-
     tion under the provisions of this Act would
     present an overwhelming and overriding risk
     to man.

The Federal government publishes lists of species that
are considered threatened and endangered for various regions
of the country. These lists appear in the *Federal Register*
(see Selected References), and are updated frequently. Both
plants and animals are included. In addition to the Federal
lists, individual states have their own official and (often
unofficial) lists of species to be protected. A source of
confusion is the fact that states often use the term "rare
and endangered species", even though this term is now
avoided by the Federal agencies.

The practical implications of this information are that
all major engineering projects must now consider threatened
and endangered (and rare) species. That is one reason why
the inventory of regional biota is useful. Moreover, it
has become necessary to assess the probable occurrence of
threatened and endangered species, even though they have
not been seen in actual field surveys. Knowledge of habitat requirements for specific species helps to assure that
such assessments are conducted in an objective manner.

It is appropriate to add that major projects actually
under construction have been stopped because of potential
harm to threatened and endangered species. A specific
case will be cited in Chapter 19.

POTENTIAL IMPACTS

Terrestrial ecosystems may be influenced by a number of natural and human-caused events. Consider, for example, the following lists:

Natural Events	*Human-caused Events*
Fires	Lumbering
Grazing	Cultivation
Hurricanes	Grazing (domestic animals)
Volcanoes	Mining
Floods	Water resources development
Droughts	Highway development
Plant disease	Urbanization

As natural events of catastrophic proportions occur, the terrain may be altered or even denuded of vegetation. The natural events leading to ecological succession will then follow, and after sufficient time (years), the system tends to go back to the climax community.

In a similar manner, humans may conduct certain events which will alter a terrestrial system. The more extensive the events, the larger the potential *impact*. Impact is used here in the sense of being synonymous with "effect" or "consequence". Impact does not mean an adverse consequence of a particular action, although it is sometimes used erroneously in that sense in environmental studies. Impacts may be specified as either "positive" or "negative" as is appropriate. The potential impacts from large-scale human activities (e.g., agriculture, highways) can be significant to terrestrial systems. Some of these will be reviewed in subsequent chapters.

Chapter 6

The Freshwater Environment

In this chapter we shift our attention to the aquatic environment. For practical reasons it is convenient to divide the aquatic environment into the freshwater environment and the marine environment. Each of these will be discussed separately in individual chapters, although many of the concepts and terms used in both aquatic environments are similar.

Organisms living in water have a different physical environment as compared with terrestrial biota. Major environmental variables for the two types of environments are summarized in Table 2. Some of these environmental factors have direct applications in environmental assessments.

AQUATIC ECOSYSTEMS

Producers in aquatic ecosystems are composed of tiny microscopic plants called *phytoplankton,* and macroscopic plants called *macrophyton*. Among the phytoplankton are various phyla of simple plants including blue-green algae, green algae, and diatoms (see Appendix 1). The macrophyton may consist of two components. One group includes the algae that are large and conspicuous to the eye; these are called *macroalgae*. Numerous species of blue-green and green algae are macroalgae. The other group consists of higher plants among the *Embryophyta* (see Appendix 1). Common examples include familiar plants such as cattails, pond lilies, and pickerel weed. Plants such as pickerel weed that grow above the water surface are often called *emergent vegetation*. The physical and chemical conditions of the water body determine which species of producers will grow.

TABLE 2. *Comparison of Physical Environmental Variables for the Terrestrial and Aquatic Environments.*

Physical environmental variable	Terrestrial environment	Aquatic environment
Supply of water	Varies greatly; may be limiting	Plentiful
Supply of oxygen	Plentiful; 21% of atmosphere	Varies; frequently a limiting factor
Sunlight	Generally plentiful	Varies; frequently limited, especially with depth
Temperature	Varies greatly over wide range	Generally restricted to narrower range
Nature of substrate	Generally stable	Varies greatly; frequently unstable
Effects of gravity	Biota adapted to gravitational stress	Medium supports; specific gravity of water similar to that of biomass

Among the consumers in aquatic systems are the relatively small animals called *zooplankton*. Most of the zooplankton are microscopic in size, although some are visible to the naked eye. Various rotifers and arthropods (see Appendix 1) are common components of the zooplankton. Biota that live on the bottom are called *benthos*. Benthic organisms include both microscopic and macroscopic species. One group that is studied commonly is called *benthic macroinvertebrates*. Many of the invertebrate phyla (see Appendix 1) are represented in the benthic macroinvertebrates, including various worms, molluscs, and arthropods. Finally, important consumers include vertebrate species belonging to various classes such as fishes, amphibians, and reptiles. Aquatic organisms which are active swimmers are called *nekton*. Fishes are the most common nektonic organisms. See Figure 5 for a diagram of an aquatic ecosystem.

Detritus forms an important component of aquatic ecosystems. Some of this organic material may be of an *autochthonous* source, i.e., generated from within the system. Dead algal debris and dead animals contribute to autochthonous materials. Some of the material may also be *allochthonous,* i.e., generated outside the water body. The deposition of dead leaves from trees into a river or lake would be an example of allochthonous inputs. Very often in freshwater systems the allochthonous source of energy (detritus) is more important than the energy input from aquatic producers.

As in all ecosystems, decomposers play an important role in the total aquatic community. Various species of fungi and bacteria break down the various organic materials on the bottom and suspended in the water column.

In summary, we can list the components of the aquatic ecosystem as follows:

Components	*Examples*
Abiotic elements	Physical conditions (temperature, light)
	Chemical nutrients (nitrogen, phosphorus, carbon)
	Substrate (sand, gravel, silt)
Producers (aquatic flora)	Phytoplankton
	Macrophyton (including macro-algae and emergent vegetation)

Components	Examples
Consumers	Zooplankton
(aquatic fauna)	Benthos (including benthic macroinvertebrates)
	Fishes
	Amphibians
	Reptiles
Decomposers	Fungi
	Bacteria

Often these categories are used in a compilation of biota for an environmental assessment of an engineering project. The precise format of an inventory will vary, but the major components generally include the groups outlined above. Of practical importance is the relative emphasis on fish species in most studies. Finally, a list of species allows an evaluation of the actual or probable presence of threatened and endangered species in the aquatic environment.

LACUSTRINE SYSTEMS

Among the freshwater systems of interest in environmental studies are various bodies of water with little or no appreciable flow. Such bodies are called *lacustrine systems,* and include ponds, lakes, and impoundments such as reservoirs. Standing water is also referred to as a *lentic* system. The physical, chemical and biological characteristics of lacustrine systems tend to vary considerably. However, all lacustrine systems have similar ecological components. What will vary are the individual species and the population densities of various biota. Relatively deep lakes tend to undergo *stratification* in the summer, with a warmer upper layer above the relatively cooler bottom layer. These layers are relatively stable and hence resist mixing. Consequently, oxygen tends to be depleted from the lower layers because of the respiration of various biota. The more biological activity in these bottom layers, the greater is the depletion of oxygen. This depletion has important implications to the biota in the system.

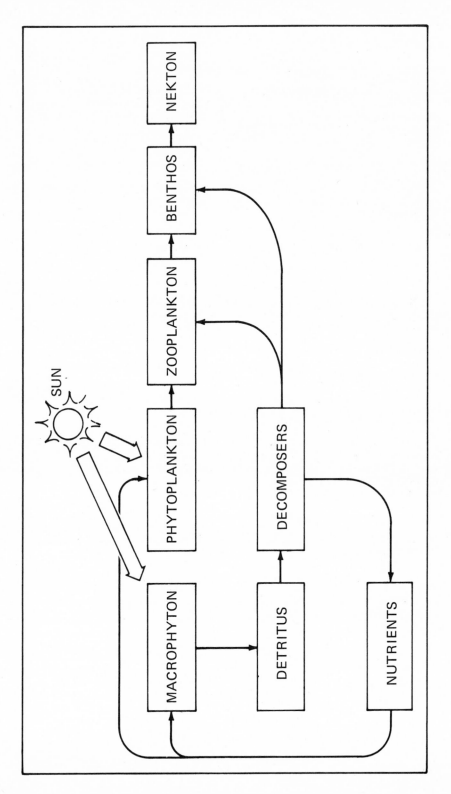

FIGURE 5.   *Diagram of an Aquatic Ecoystem.*

RIVERINE SYSTEMS

Riverine systems, in contrast with lacustrine systems, have a steady flow of water. Such systems with running water are also called *lotic* systems. This flow of water is an important characteristic of the riverine system, and it determines the biota to a great degree. The reaches of a river or stream with a steep gradient will have rapid flow. These areas are often scoured and characterized by turbulent flow or even rapids. These are called *riffles*. Other areas with low gradients may have very slow flow or even relatively stagnant pools. The composition of the stream bed or substrate is another important characteristic. Finally, the amount of energy, both from primary producers and allochthonous sources, is also important. Riffle areas tend to have high oxygen levels because of turbulent mixing. In contrast, stagnant reaches may have low oxygen levels, especially in summer. Producers, consumers, and decomposers are present in all riverine ecosystems.

THE RIPARIAN HABITAT

The interface between the terrestrial environment and the aquatic environment is often a location of interest to biologists. This area forms a habitat called the *riparian habitat,* and many species may be present. Muskrats, raccoons, mink, otter, amphibians, and various birds of prey will use this habitat. Extensive vegetation along the bank will enhance the probability that the area will support various forms of wildlife. Frequently these areas are biologically productive areas, and consequently they receive attention from fish and wildlife agencies. This is especially true in areas with extensive agricultural land, such as in the midwest. The riparian zone can provide habitat diversity between the aquatic environment and the open fields. The transition between an aquatic system and a terrestrial system is often unclear; thus the limits of the riparian zone are frequently arbitrary. It is useful to remember that all systems are ultimately linked by complex food webs and biogeochemical cycles.

TROPHIC CONDITIONS

The *trophic state* of a body of water is of important, practical significance to environmental studies. A body of water with few nutrients and low standing crop of algae is called *oligotrophic* (few or little nourishment). On the other hand, a lake with high levels of nutrients and a dense algal community is called *eutrophic* (true or good nourishment). There is a natural process which occurs in lakes and ponds called *eutrophication,* in which nutrients increase with time. Sediment accumulates, nutrients increase, macrophytes increase, and with the passage of time the pond may actually fill in. The process of eutrophication may be accelerated by human-caused sources of inputs, such as sewage input or runoff from agricultural land. The trophic condition of a lake or pond has an important bearing on the water quality and the nature of the biological community. See Table 3 for a comparison of oligotrophic and eutrophic lakes.

WATER QUALITY

Water quality is another important condition in the aquatic environment. Water quality is an integrative concept, and consists of an assessment of the condition of the water in relation to some goal. Consequently, it is necessary to define *criteria* for the evaluation of water quality. Since there are several hundred parameters (e.g., temperature, color, bacteria) that may be measured in water, selection of specific parameters are usually among the criteria for the assessment of water quality. For aquatic biology, there are certain water quality parameters that are important; some of these are compared in Table 3. It is important to note that there is no single, numerical index for water quality. Consequently, water quality is frequently classified on a non-quantitative scale (e.g., A,B,C), although individual criteria may be assigned numerical values. Sometimes specific goals for water quality are established, and individual criteria are assigned numerical values consistent with such goals. Such a value then becomes a *standard* required to achieve the goal.

TABLE 3. *Comparison of Selected Properties and Parameters for Oligotrophic and Eutrophic Lakes; (ppm indicates parts per million).*

Selected properties and parameters	Oligotrophic lake	Eutrophic lake
Biological productivity	Low productivity	High productivity
Oxygen	Abundant; usually above 5 ppm	May be lower or even depleted at bottom, below 1 ppm
Nitrogen	Low concentration; usually below 0.1 ppm	High concentration; usually above 0.1 ppm
Phosphorus	Low concentration; usually below 0.01 ppm	High concentration; usually above 0.01 ppm
Dissolved solids	Low concentrations	High concentrations
Organic materials	Low concentrations	High concentrations
Color	Low to absent	Moderate to high
Depth	Usually deep	Usually shallow

GROUNDWATER

Surface waters are only part of the freshwater supply. A very large additional supply exists in the groundwater. Groundwater is an important source of water supply in many parts of the country, especially as some surface waters have become contaminated. The natural constituents in groundwater vary considerably, and depend upon the mineral content and organic matter in various strata of soil and bedrock. The quality of groundwater is important to humans when the water is used as a drinking water supply. Groundwater contamination from human activities has been recognized as a widespread problem. Also, various engineering projects can influence the quantity of groundwater in localized areas. Groundwater is now receiving considerable attention in environmental studies.

PUBLIC HEALTH

The importance of water quality to human health has been recognized for over 100 years. The public health profession is concerned with the maintenance of acceptable water quality standards for drinking water, water used for contact recreation, and water from which fish for consumption are derived. Certain water quality parameters receive constant attention, including bacteria, heavy metals, and various organic chemicals. In recent years, advances in technology have resulted in the additions of numerous chemicals to surface waters and to groundwater. These chemicals may influence aquatic biota and may also have important public health implications. More will be said of these potential problems in subsequent chapters.

POTENTIAL IMPACTS

The freshwater environment experiences many dynamic changes induced by various natural events. Hydrologic cycles, fluctuations in river stage, seasonal changes in temperature and solar radiation, and watershed runoff all contribute to these changes. Aquatic biota must survive the impacts of these natural events.

Against this background of natural events, people may contribute additional physical and chemical impacts. The discharge of sewage, runoff from agricultural lands, runoff from urbanized areas, erosion from construction activities, thermal effluent from electric generating stations, and accidental spillage of toxic chemicals may all have significant impacts on aquatic systems. Many of these impacts have been the historical bases for the enactment of the legislation discussed in Chapter 2. One of the major challenges facing environmental biologists and engineers is to assess these impacts in an objective manner.

Chapter 7

The Marine Environment

The marine environment presents some dramatic differences when compared with the freshwater aquatic environment. Of primary importance is the relative vastness of the oceans; over 70% of the earth is covered with sea water. The average depth of the North Atlantic is about 6000 meters (3.8 miles). The oceans are linked dynamically by vast circulation patterns; consequently it is difficult to delineate relatively well-defined marine ecosystems. Finally, the oceans are the ultimate chemical sink for all substances originating on the land masses and flowing with water to the sea. These substances in solution add to the characteristic salt content of sea water.

SOME IMPORTANT ABIOTIC FACTORS

One of the more important properties of the marine environment is the salt content or *salinity*. The five most abundant chemical constituents in solution in sea water are as follows:

Constituent	Concentration in ppm
chloride	19,350
sodium	10,770
sulfate	2,700
magnesium	1,300
calcium	400

These values will vary slightly at different locations. However, it is of interest to compare this chloride level with typical surface waters of the North American continent (approximately 10 ppm). The concentrations of salts are so high in sea water that it is more customary to talk in terms of parts per thousand (°/oo). Thus chloride values would be 19.3 °/oo for ocean water. The total salinity of sea water is about 35 °/oo. Thus the salinity of the ocean is about three orders of magnitude (1000 x) greater than that of freshwater. As with all natural waters, thousands of other chemical species (inorganic and organic) are also present in sea water.

Another important characteristic of the oceans is the nature of the *temperature profile*. As shown in Figure 6, temperature decreases with depth. The values shown are characteristic of the North Atlantic at approximately mid-latitude. Notice the rapid decrease in temperature in the region of the *permanent thermocline*. The deeper layers of the ocean have a temperature of only a few degrees Celsius. The upper layers between 15 and 20 degrees Celsius have enormous amounts of thermal energy. These layers of rela-tively warm water flow in horizontal currents and set up global circulation patterns.

Sea water is constantly experiencing dynamic movements such as:

waves   horizontal currents
tides   vertical currents

These currents are important because they result in the transport of nutrients, oxygen and various biota. The movement of planktonic larvae across thousands of miles of ocean is not unusual. Vertical currents rising up from nutrient-rich bottom water (upwelling) often result in high biological productivity in localized areas. Wave and tidal movements act as pumping mechanisms to circulate water in and out of coastal zones and estuaries. Many marine organisms are adapted to these dynamic events.

Materials in suspension are constantly being carried by rivers to the ocean. Mechanical forces cause coastal erosion and contribute suspended materials. Finally, the bodies of dead organisms, especially microscopic forms, become suspended in the water column. Eventually much of this material settles out to the bottom, forming deposits

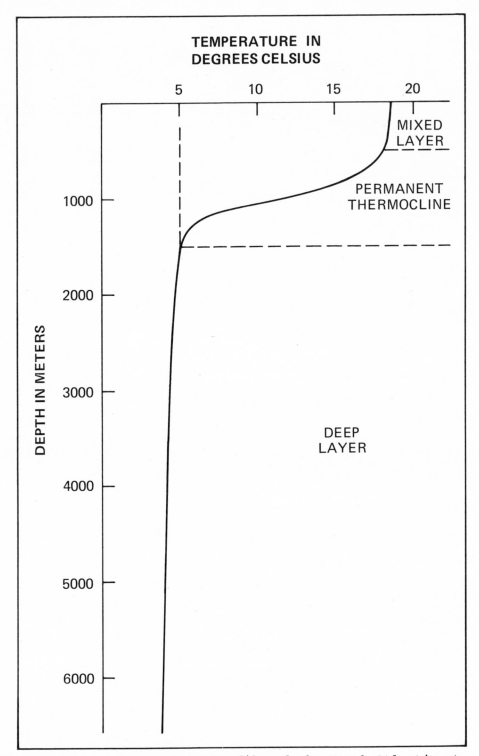

FIGURE 6.   *Temperature Profile of the North Atlantic at Approximately Mid-Latitude.*

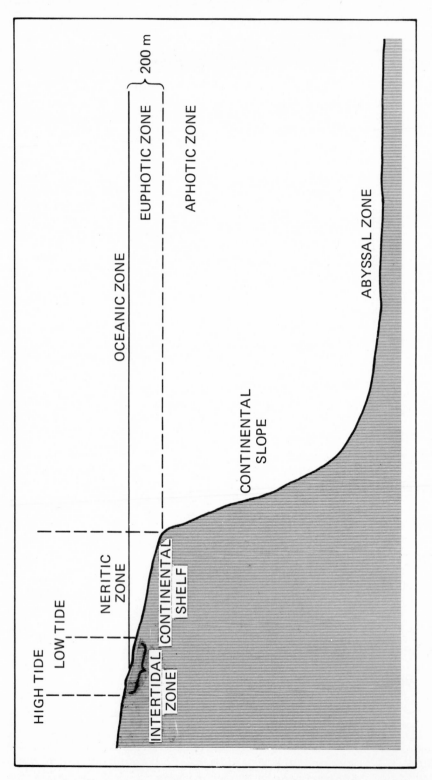

FIGURE 7.   *Diagram of a Profile of the Marine Environment from the Coastal Area to the Mid-Ocean Area.*

of *sediments*. Layers of sediments, deposited over the
geologic ages, are characteristic of the bottom of the
ocean. Suspended particles can increase the turbidity
and limit light penetration, especially off the coast.
The discharge of the Amazon River forms a dramatic turbidity zone many miles at sea. Similar phenomena are familiar sights to mariners throughout the world.

ZONATION

The marine environment has a number of zones, and a
definition of these zones makes the marine biota easier
to understand. Figure 7 is a profile of the marine environment from the coast to deep oceanic areas. Note that the
classification of zones is based upon:

    vertical relief of the bottom
    horizontal regions
    light penetration

The zone between high and low tide is called the *intertidal
zone;* it is also called the *littoral zone*. There is a
continental shelf that extends many miles out from the
coast. The continental zone usually extends to a depth of
200 meters (100 fathoms). Beyond the shelf there is
another descending gradient that extends to the bottom of
the ocean basin. This gradient is called the *continental
slope*. The bottom of the ocean is called the *abyssal zone*.
    The open water mass above the continental shelf is
called the *neritic zone*. Beyond the continental shelf the
open water is referred to as the *oceanic zone*. The *euphotic zone* is the zone where light penetration and hence
photosynthesis occurs. Below that depth, in the *aphotic
zone,* there is no penetration of light. The actual depth
of the euphotic zone varies considerably from region to
region because of the turbidity of the water. It becomes
apparent that only a comparatively thin upper layer of the
ocean is occupied by primary producers.

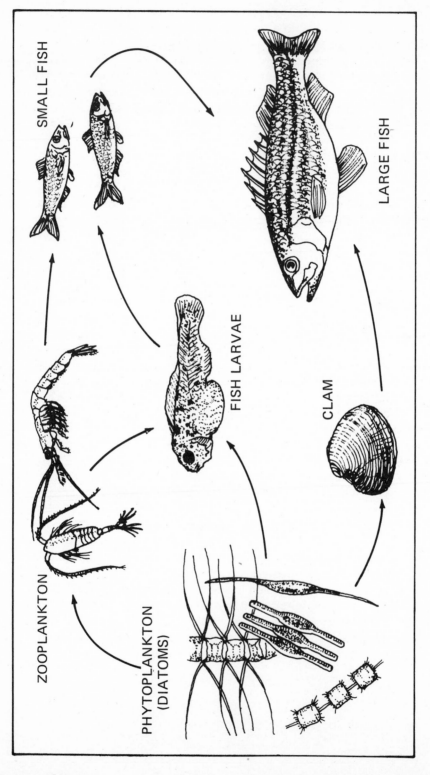

FIGURE 8.   Diagram of a Simple Marine Food Web.

MARINE BIOTA

Producers in the marine environment include both macrophytic and planktonic algae. However, the vast majority of the energy derived in the open sea is from planktonic algae, especially diatoms. Consumers include zooplankton of many types, especially various crustaceans. The benthos of the marine environment includes many types of animals such as worms, molluscs, starfish and crustaceans. Benthic animals have been found in the deepest parts of the ocean, at depths over 10,000 meters. The oceans also include numerous species of *nekton*. Nekton are animals that are powerful enough to swim and thus are not passively carried by ocean currents as are the plankton. Among the nekton are various types of molluscs, crustaceans, and fishes. Finally, as with all ecosystems, the oceans also contain detritus and decomposers. In summary, the biota of the marine environment include the following:

Components	*Examples*
Producers (marine flora)	Macrophyton (especially green, brown and red algae)
	Phytoplankton (especially diatoms)
Consumers (marine fauna)	Zooplankton (especially crustaceans)
	Benthic invertebrates (including various worms, molluscs, crustaceans and echinoderms)
	Nekton (including larger molluscs, crustaceans, and fishes)
Decomposers	Fungi
	Bacteria

The actual species composition will vary greatly with abiotic conditions, latitude, and habitat. Figure 8 illustrates a generalized marine food web.

MARINE HABITATS

The marine environment presents an interesting diversity of habitats for marine biota. The *pelagic* habitat
consists of the open sea, and includes many types of planktonic and nektonic species. Closer inshore along the
coast and along islands are a number of other habitats such
as *mud flats, sandy beaches,* and *tidal pools*. A rich
variety of macrophytic algae and invertebrates may be
found in these areas. In warmer latitudes *coral reefs* and
mangroves offer special habitats for a variety of marine
life. Stretching along the coast for thousands of miles
is an extensive network of *salt marshes*. From an ecological perspective, these salt marshes are important because
they offer habitat for waterfowl, they carry on extensive
primary productivity, they provide nursery grounds for
young fish, and they support a diversity of invertebrates.
Abiotic factors such as the geology, salinity, tidal currents and temperature are important in determining the
nature of these marine habitats. A special case of practical importance is the estuary; this will be discussed in
the next section.

ESTUARIES

An *estuary* is a semi-isolated body of water near the
coast separated from the open sea by a partial barrier.
Usually the estuary is influenced both by fresh water and
by tidal currents. Consequently, the salinity of estuaries
is below 35 ^{o}/oo, but above that of fresh water. This
mixture of sea water and fresh water is called *brackish
water*. Geologic details for estuaries vary greatly.
Estuaries are generally close to shore and hence are
not deep. Therefore there is ample light penetration for
photosynthesis. Terrestrial runoff usually brings rich
organic materials and nutrients. Tidal currents provide
dynamic mixing actions. Surface waters are generally warm,
and hence the waters of estuaries are relatively warm.
Many of these factors often combine to produce a high density and a rich diversity of estuarine biota. Thus estuaries are typically areas of high biological productivity.

However, estuaries may also be regions of intense human activity. Coastal zones are often heavily populated. Historically, river deltas and mouths are areas of early urbanization and industrial development. Rivers are also areas of shipping and navigation. Estuaries are usually the first marine areas to experience the effects of discharge of waste materials. There various human activities, therefore, may interact with coastal areas, including estuaries. Sometimes, there interactions will cause impacts on coastal ecosystems.

POTENTIAL IMPACTS

Coastal zones and offshore areas have many economic implications for humans. From early times marine fisheries have been important food sources. Today some of the most productive commercial fisheries are still over the continental shelf. The shelf has also taken on additional importance in recent years because of offshore drilling for gas and oil. Extensive shipping lanes follow coastal areas. Coastal areas are also receiving increased attention as potential sites for large electric generating stations. Finally, physical changes of the coast are frequently necessary with such engineering projects as breakwaters, canals, land reclamation, and dredging.

These human activities have various short term and long term implications to the marine biota. Overfishing can have long term impacts on population dynamics and future fisheries. The introduction of hydrocarbon chemicals may have both short term and long term impacts on the health of the aquatic community. Thermal discharge of cooling water from electric generating plants may have both beneficial and detrimental impacts on the coastal ecosystem. Finally, marine engineering and navigation projects may influence turbidity, circulation patterns, and temperature profiles. These changes in abiotic conditions in turn may have impacts on the marine biota.

Areas with intense human activity, such as coastal areas, will ultimately be subject to conflicting objectives. With optimal planning it is possible to achieve multiple objectives; it will never be possible to satisfy all objectives. Among the objectives mandated by law today are those dealing with the protection and conservation of marine resources (see Chapter 2). It is possible to meet these conservation objectives and still maintain some of the traditional objectives of human productivity.

Chapter 8

Wetlands

Wetlands comprise unique habitats which include many of the features of both aquatic and terrestrial ecosystems. Much attention has been focused on wetlands in recent years. Also, current legislation puts heavy emphasis on the protection of wetlands.

HISTORICAL OVERVIEW

Public interest in wetlands in the United States has existed for well over 100 years. Many years before the recent legislation (see Chapter 2) various laws were passed on wetlands in the United States.

Interest in the nineteenth century centered on the possible reclamation of wetlands for human settlement. The *Swamp Land Acts of 1849, 1850 and 1860* were enacted to encourage a program of reclamation. Flood control and the elimination of mosquito-breeding areas in the interest of public health were additional objectives. This program succeeded and eventually millions of acres were reclaimed and taken over for private interests.

At the turn of the century various groups felt that conservation of wetlands was an important national objective. The U.S. Department of Agriculture became interested in the agricultural potential of wetlands, and conducted inventories of wetlands in 1906 and 1922. Diverse interests centering on development, flood control, agriculture, conservation and waterfowl habitat developed during

TABLE 4. *Classification of Wetlands and Deepwater Habitats (Cowardin et al., 1979).*

Systems	Subsystems	Classes	
Marine	*Subtidal*	*Rock bottom*	*Reef*
	Intertidal	*Unconsolidated bottom*	*Rocky shore*
		Aquatic bed	*Unconsolidated shore*
Estuarine	*Subtidal*	*Rock bottom*	*Rocky shore*
	Intertidal	*Unconsolidated bottom*	*Unconsolidated shore*
		Aquatic bed	*Emergent wetland*
		Reef	*Scrub-shrub wetland*
		Streambed	*Forested wetland*
Riverine	*Tidal*	*Rock bottom*	*Unconsolidated shore*
	Lower perennial	*Unconsolidated bottom*	*Emergent wetland*
	Upper perennial	*Aquatic bed*	*Streambed*
	Intermittent	*Rocky shore*	
Lacustrine	*Limnetic*	*Rock bottom*	*Rocky shore*
	Littoral	*Unconsolidated bottom*	*Unconsolidated shore*
		Aquatic bed	*Emergent wetland*
Palustrine		*Rock bottom*	*Moss-lichen wetland*
		Unconsolidated bottom	*Emergent wetland*
		Aquatic bed	*Scrub-shrub wetland*
		Unconsolidated shore	*Forested wetland*

this period. Following World War II the rapid development
of urbanized areas, especially in coastal areas and river
basins, resulted in the draining and filling of extensive
areas of wetlands.

In 1956 the Fish and Wildlife Service, U.S. Department
of the Interior, published the results of an extensive
survey of wetlands in the United States *(Circular 39)*.
The main focus of this survey was wetlands as habitat for
waterfowl and other wildlife. Subsequently, national
interest in conservation and environmental protection has
lead to the enactment of various laws whose provisions
include the protection of wetlands.

CLASSIFICATION OF WETLANDS

For twenty years, *Circular 39* (Shaw and Fredine, 1956)
served as "the Bible" on wetlands, and it has been an
extremely useful publication. Because of increasing
interest, it was reissued in 1971 without change. In this
circular, 20 wetland types were described, and the classification system was accepted as the standard system by
professionals interested in fish and wildlife resources.

During the 1970s it became apparent that *Circular 39*
was no longer adequate for dealing with environmental
issues involving wetlands. Whereas the central focus of
the Fish and Wildlife Service survey was wildlife habitat,
a much wider spectrum of interests and needs has developed
recently. In 1975, a National Wetland Classification and
Inventory Workshop was sponsored by the Office of Biological
Services, Fish and Wildlife Service. Representatives from
Federal, Canadian, and state agencies, along with private
organizations, were present. An *Interim Classification
of Wetlands and Aquatic Habitats of the United States*
was published in the proceedings (Sather, 1976). An
operational draft on *Classification of Wetlands and Deepwater Habitats of the United States* was issued in 1977.
Finally, a revised version of this important report was
published in 1979 (Cowardin *et al.,* 1979). This new
classification system for wetlands and deepwater habitats
is summarized in Table 4.

The new classification system also considers other
attributes of wetland systems, such as water chemistry,
inundation cycles, flow regime, geologic features, and
dominant biota. Details are beyond the scope of this book,
and the interested reader should consult Cowardin *et al*.
Another useful feature of this publication is the comparison
of the new classification system with that in *Circular 39*
(Shaw and Fredine, 1956).

IMPORTANCE OF WETLANDS

Wetlands are often highly productive systems. Coastal
marshes are among the more productive ecosystems known.
Primary producers supply energy to nearby estuaries and
coastal areas. Detritus forms a rich energy source for the
estuarine food web. Juvenile fish of such species as the
menhaden use the marshes as nursery grounds, feeding on
detritus. Some inland wetlands are also productive systems.
Thus wetlands may be viewed as sources of energy for the
diverse biosystems (consumers) that eventually depend upon
them.
Wetlands also provide habitat for diverse species of
organisms. Many invertebrates such as snails, bivalve molluscs, crustacea, and insects are found in various wetlands.
Vertebrates may include amphibians, reptiles, and birds.
Riparian wetlands near the shore in riverine systems often
provide habitat for numerous birds and mammals. Wetlands
of coastal areas also have a rich variety of animal species.
Wetlands often provide specialized and unique habitats, and
therefore are important from an ecological perspective.
Often wetlands provide high species diversity in an other-
wise uniform landscape.
The importance of wetlands has been recognized in
Executive Order 11990 (see Chapter 2). Federal agencies
are now developing guidelines on the evaluation and protection of wetlands. Recent examples include documents
from the U.S. Army Corps of Engineers (Reppert *et al.*, 1979)
and the Federal Highway Administration (Erickson *et al.*,
1980). Finally, the American Water Resources Association
has published a major report on a National Symposium on
Wetlands (Greeson *et al.*,1978). This document summarizes
the current state of our knowledge on wetland functions and
values.

FLYWAYS

A dramatic example of habitat may be seen in the utilization of wetlands by waterfowl. Migratory waterfowl fly north during the spring and south during the fall. Four major *flyways* (Atlantic, Mississippi, Central and Pacific) are used by these waterfowl (see Figure 9). Millions of acres of wetlands along these flyways provide food and habitat for millions of migratory waterfowl. There is much interest by the Fish and Wildlife Service in these flyways, and they receive much support from thousands of sportspersons.

PUBLIC INTEREST

During the past decade public interest in wetlands has intensified greatly. A great diversity of individuals and organizations with a variety of objectives often become interested in any project which may encroach on wetlands. Conservation groups and environmental groups may oppose any potential loss of wetlands. Often they will resort to adjudication, using a variety of legal mechanisms to support their position. Equally firm conservation positions may be taken by individual or organized sportspersons. Since the funds they provide help to maintain the fish and wildlife management programs, their positions are taken seriously. In addition to these policy or special-interest positions, there is the position that may be taken on scientific grounds. Wetlands may represent unique habitats of high biological productivity and species diversity. Consequently, they demand careful assessment on technical grounds. Finally, the general public is aware of the high public profile of wetlands. These various factors combine to make wetland areas of special concern for many engineering projects. Potential impacts must be assessed carefully, and mitigation measures for adverse impacts will generally be required.

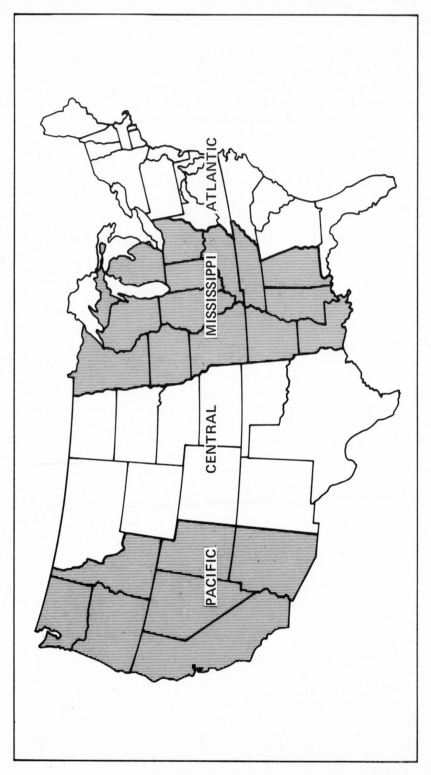

FIGURE 9.   *Four Major Flyways of the United States.*
(From Circular 39; Shaw and Fredine, 1956)

POTENTIAL IMPACTS

The potential impacts of various projects on wetlands have been documented extensively. Short term impacts from construction may include the following:

> disruption of primary producers
> disruption of benthic organisms
> disruptions of waterfowl
> changes in hydrology and hydrodynamics
> changes in physical - chemical properties of
>   the water column

These changes may be caused by cut and fill operations, drainage ditches, dams, dredge and disposal operations, and the construction of marine facilities.

Longer term impacts are also possible. Drainage and/or filling in of wetlands, of course, results in the loss of the wetlands as habitat. Projects may cause long term modification of the hydrology and the margins of wetlands even though there is no direct development in the wetlands themselves. The operation and maintenance of projects adjacent to wetlands may result in the introduction of potentially toxic substances such as salts, pesticides, and petroleum hydrocarbons. Disposal of solid waste may also cause impacts on wetlands.

Wetlands and the often-associated flood plains and coastal plains are now receiving careful attention from various planning agencies. Projects are usually coordinated with such agencies. Through effective liaison with wildlife management agencies, planning agencies, and, where appropriate, with citizen groups, many of the problems associated with project development near wetlands will be avoided.

Chapter 9

Fish and Wildlife

HISTORICAL OVERVIEW

Another area of public interest is the fish and wildlife resources of the region. Interest in conservation and
the management of wildlife began at the end of the nineteenth century. State and Federal agencies were established
to restore and conserve the Nation's wildlife resources.
Sportspersons supported these efforts, and for many decades
professional fish and wildlife biologists have cooperated
with various sports organizations. Consequently, there is
a historical interest by fish and wildlife agencies in
species of interest to sportspersons.

Government involvement in the protection and promotion
of commercial fisheries also began in the nineteenth century. Federal laboratories were soon established to conduct research in fisheries, including the development
of technology. Primary emphasis was in shellfish (molluscs
and crustaceans) and in finfish (vertebrates). The economic importance of both freshwater and marine commercial
fishing grew steadily. However, there were subsequent
declines in commercial fisheries in some areas, mainly due
to water quality changes.

After World War II interest in conservation, recreation,
and pollution control steadily increased. Various laws for
the protection of wildlife and wildlife habitat were enacted (see Chapter 2). Important among these laws were the
Fish and Wildlife Act of 1956 and the *Fish and Wildlife
Coordination Act of 1958*. In 1967 the Secretary of the
Interior and the Secretary of the Army signed a *Memorandum
of Understanding,* agreeing that certain engineering projects which involved potential impairment of fish and

wildlife resources would receive careful evaluation. The
National Environmental Policy Act of 1969 (NEPA) and the
Federal Water Pollution Control Act Amendments of 1972
added additional impetus to the protection of wildlife.
   In the implementation of environmental assessment
and protection, most agencies through policy and practical
experience focus their interests on fish and wildlife.
The historical interests shown by natural resources agencies, by sportspersons, and by the general public promote
such policies. The consequence of these historical events is
that environmental assessments which involve biological
resources will predictably focus on certain fish and wildlife
species.

FISHERIES BIOLOGY

   All states of the United States and the Federal government support various agencies with professional fisheries
biologists. Among their activities are those involving
habitat assessment. They are frequently concerned with
stream surveys and lake surveys. Such physical and chemical features as the nature of the bottom, water quality,
erosion, vegetation, spawning areas, and sources of food
are often studied. Thus they evaluate many of the abiotic
and biotic conditions of the water. They also study the
species diversity and the *density* of the fish populations.
From these studies the status of fish populations in various
bodies of water are determined.

FISHERIES MANAGEMENT

   Most fish and wildlife agencies, however, have specific
objectives mandated by the legislative acts that created
them. These objectives include the management of fisheries
so that there are sustained yields of fish for recreation
and commercial fisheries. Various management activities
include *fish control, stocking,* taking *creel census,* and
studying populations by *year class* to determine the
success of annual spawn.

Fisheries are often divided into *warmwater species* and *coldwater species*. Coldwater species are highly prized by sportspersons, and include such species as trout and salmon. Common warmwater species include the largemouth bass, catfish, gizzard shad and sunfish. Intermediate between the coldwater fish and the warmwater fish are such species as the smallmouth bass, walleye, and the northern pike. In addition to temperature, other water quality parameters such as dissolved oxygen and turbidity are important factors in determining the suitability of fish habitat.

The *carrying capacity* is a term commonly used by fish and wildlife biologists. In the aquatic environment it refers to the maximum poundage of fish that a given area of specific habitat can sustain at any one time. Many environmental factors will influence the carrying capacity. The carrying capacity for temperate lakes in the United States may sustain a *standing crop* of several 100 pounds of fish per acre. *Production* refers to the additional weight of fish added during a specific time period. Sometimes the term *annual yield* is used to denote the production added on an annual basis. The annual yield may be quite large, often as high as 50-100 percent of the standing crop. A 100 percent yield, for example, means that half of the standing crop may be harvested without endangering the capacity for sustained yields.

WILDLIFE BIOLOGY

As mentioned previously in Chapter 5, wildlife refers to all animals, although common usage often restricts the term to the larger terrestrial animals. In a parallel sense with fisheries biology, various agencies also employ terrestrial wildlife biologists. *Habitat assessment* is again a major professional activity. Habitat for animals depends upon vegetation, and thus habitat assessment includes the evaluation of *overstory* vegetation (trees) and *understory* vegetation (shrubs, herbs, grasses). Frequently professional foresters and botanists are involved in vegetation surveys. Such features as cover, food and water are studied. Wildlife biologists are also involved in animal surveys. Various techniques for taking census are used. Thus the *population dynamics* of certain species of birds and mammals are studied. See Appendix 3 for additional comments.

WILDLIFE MANAGEMENT

Wildlife management centers its activities on the management of habitat so that wildlife will be protected and allowed to reproduce. A closely-related profession is *game management,* which includes in its objectives the achievement of sustained annual yield of wild game for recreational use. Measures to control and manage game began in the Middle Ages in Europe. In colonial times most of the colonies enacted closed seasons to protect certain game species. Game management has since evolved into a highly-developed science. Today most wildlife agencies include both *game species* and *non-game species* within their sphere of interest. However, the historical interest in game species is very strong, and such species will receive close evaluation in environmental assessments.

The concept of carrying capacity defined under aquatic habitats can also be applied to terrestrial habitats. Carrying capacity is the number of individuals of a certain species that a given area of habitat may sustain. Again it becomes apparent that the carrying capacity of an area will depend upon the vegetation and other environmental features. Deer may require 10-50 acres per animal. On the other hand, a small species such as the cottontail rabbit may have a density of 0.5 animals per acre. Game birds such as grouse and quail will also fluctuate widely in density. However, a figure of one bird per acre is a good approximation. It is important to realize that these numbers are intended only as indications of orders of magnitude. Specific locations vary greatly, and local wildlife biologists should be consulted for estimates of carrying capacity.

Two factors combine to make the study of animal populations (census) extremely difficult. All wild animal populations undergo dynamic fluctuations; these may be caused by predator-prey relationships, disease, destruction of habitat and catastrophic events (fires, hurricanes). Wildlife biologists have conducted numerous studies of population dynamics, and an extensive literature exists. Another factor is the mobility of animals, especially the larger mammals such as deer, mountain sheep and wolves, and waterfowl such as ducks and geese. These animals often undergo seasonal migrations over long distances. Even within its home range (see Chapter 5), an animal will show great mobility. It often takes years of patient study by wildlife professionals to study these migration and mobility

patterns. Predictably, then, these are some of the issues
that may be raised in any engineering project that may
interact with the habitat of wildlife species.

The practical consequence of these realities is that
fish and wildlife agencies will tend to center their interests in wildlife habitat rather than specific lists or inventories of fauna. The difficulties of surveying animal
populations usually preclude accurate assessments of actual
standing crop of wildlife species. Therefore, the emphasis
is on wildlife habitat, and the potential impacts of any
project on the carrying capacity for wildlife.

CRTICAL HABITAT

Under the *Endangered Species Act of 1973* (see Chapters
2 and 5) one of the requirements is the assessment of critical habitat for threatened and endangered species. Projects
which may have adverse effects on *critical habitat* for these
species must be evaluated carefully. The concept of critical habitat centers on the fact that certain species are not
easily adaptable to generalized habitat conditions, and
thus require certain environmental conditions for their
survival and successful propagation. The presence of critical habitat does not automatically exclude any project activities in the general area. However, it does require the
careful assessment of the probable consequences to threatened and endangered species.

IMPACT ASSESSMENT

It now becomes possible to predict some of the potential
impacts of a project on fish and wildlife resources that
will have to be evaluated. One approach is to list a series
of questions that are raised frequently in environmental
assessments.

1. What will be the effects on dissolved oxygen?
2. What will be the effects of construction on
    erosion and turbidity?
3. What will be the effects on spawning areas?
4. What will be the effects on the warmwater
    sports fishery?

5. What will be the effects on the coldwater sports fishery?
6. What will be the effects on the carrying capacity for terrestrial wildlife species?
7. What will be the effects on animal migration?
8. What will be the effects on breeding and nesting behavior?
9. What will be the effects on critical habitat?
10. What will be the effects on threatened and endangered species?

The answers to these questions are seldom simple and straightforward. Fish and wildlife populations undergo dynamic changes even without a proposed project in the area. They are constantly interacting with the physical environment. For these reasons an appreciation of the dynamic events in ecosystems is necessary for predicting changes to any ecosystem. Only then can the answers to these and other questions be attempted in an objective and scientific manner.

Chapter 10

Environmental Stress

The concept of environmental stress is sufficiently important to merit a separate discussion. Many of our current studies in environmental biology, especially in relation to engineering projects, must consider the concept of stress. Unfortunately, this subject, as a coherent area of specialized research, has not received the attention it needs. However, much of the theory base for the analysis of environmental stress has been in existence for many years. As examples, the analysis of biological organisms in relation to various salt concentrations in the environment, to various oxygen tensions, to various changes in temperature, and to ambient light conditions has been going on for almost one hundred years. General physiologists, comparative physiologists, fisheries biologists, plant physiologists, geneticists and various other specialists have been working on these research problems over this time period. Consequently, a significant literature exists, even though it has not been organized in the context of environmental stress. Stress is used in this book in the sense of a force or condition exerted upon an organism or biosystem such that it tends to displace or influence that organism or biosystem away from an existing state. In that sense environmental stress is a physical concept and is not intended to imply any social value or judgment.

ADAPTATION

A fundamental attribute of living organisms is their capacity to adapt to a variety of physical conditions in the environment. The phenomenon of *adaptation* centers on the dynamic responses of organisms to dynamic changes in the

environment. With minor exceptions, the environment normally undergoes dynamic fluctuations, and many organisms
can adapt to these fluctuations.

Not all organisms have the same potential for adaptation.
For example, marine species living in estuaries may not
necessarily survive in the pelagic zone. Some plants
adapted to the conditions of the deciduous forest cannot be
expected to adapt to the desert. As another familiar example, ducks and geese adapt to climatic changes by flying
north and south for great distances. Thus the mechanism
for adaptation may include the following:

   physiological mechanisms
   structural (morphological) features
   behavioral mechanisms

Many animals can adapt to a relatively wide range of
vegetation types. Such animals typically have a large geographic range (see Chapter 5). Other animals can only
survive within certain, specialized plant communities. Some
carnivores may have specialized diet requirements, whereas
others are more adaptable in their diets. Thus adaptation,
although a fundamental attribute of living things, is also
an extremely variable phenomenon.

This variability is an expression of the *genetic traits*
inherent in each individual organism. The genetic traits
define the limits within which an organism can adapt to the
environment. Organisms also have the capacity of *acclimation,* or the ability to adjust basic mechanisms to some environmental condition. Thus certain fish may become accli-
mated to higher temperatures than what they might normally
encounter. This acclimation can only occur within the basic
limits determined by the genetic traits.

NATURAL SELECTION

Although a population is a collection of individuals of
the same species, there can be conspicuous differences among
individuals within the same species. Thus plants of a certain species within a localized area may be physically different from plants of the same species at another location.
Such differences in organisms within the same species are
called *ecotypes,* and there is strong scientific evidence

that ecotypes are genetically distinct. Thus organisms within the same species (i.e., interbreeding *gene pool)*, may have individually distinct genetic differences, and these differences may often become apparent under certain environmental conditions.

These events are manifestations of the process of *natural selection.* This process is best understood by considering the entire (or very large) population (gene pool) of a species over a long time frame. Certain environmental factors may undergo changes which place stress on individuals within a species in the area. Those individuals within the species whose capacity to adapt under the ambient conditions is greatest will tend to survive in greater numbers. The individuals whose genetic make-up results in a lesser capacity to adapt to ambient conditions will tend not to survive. Thus over a period of time the surviving generations of individuals within the species will be selected in favor of those with the appropriate adaptive mechanisms. This is the continual process of natural selection.

Natural selection also operates with populations representing more than one species. Species better adapted to particular environmental conditions will survive at the expense of others not so adapted to those conditions. Thus over a period of time, there will be a tendency for the natural selection process to favor (1) those species which can best adapt to the environmental conditions, and (2) those individuals within a species that can best adapt to the environmental conditions. Some of these concepts can be analyzed by examining the effects of environmental stress caused by abiotic factors.

ABIOTIC FACTORS IN THE ENVIRONMENT

Many of the environmental factors mentioned in Chapter 3 are important to the concept of environmental stress. In fact, it is the ability of certain organisms to survive and the inability of others to survive that determines the characteristic biota within the various biomes. However, aside from these adaptive phenomena on a very large geographic scale, it is useful to examine stress on a simpler scale.

   Figure 10 is a diagram of the survival potential of a
population of aquatic organisms under different dissolved
oxygen concentrations in the water. Note that at concentrations above 5 ppm, there is 100 percent survival. As the
dissolved oxygen is reduced, there are a number of mortali-
ties. Eventually under relatively anoxic conditions
(approximately 0 ppm) practically the entire population
will succumb. Note also the arbitrary division of the
population into three classes of individuals:

       sensitive
       intermediate
       tolerant

Should the ambient oxygen concentrations under natural conditions remain above 5 ppm, practically all organisms
would survive (unless influenced adversely by some other
factor). On the other hand, should ambient oxygen conditions remain below 2 ppm for any significant period of time,
only the tolerant individuals would survive. The tolerant
species would be selected out and would transmit their
characteristics to future, surviving generations of the
species within that system.
   The foregoing analysis illustrates the point that survival under environmental stress is a statistical concept.
As the oxygen becomes depleted, environmental stress increases (Figure 10). As stress is increased, the probability of death increases. Such a relationship is called a
stress-effect curve. We could apply the same analysis to
other abiotic factors in the environment such as temperature, moisture, wind velocity and hydrogen ion concentration (pH). The quantitative relationships will be similar.
As the environmental conditions depart from the optimal
value for the species, stress is imposed. As the stress increases, the probability of mortality increases. Finally
the limit of tolerance is reached beyond which no individual
of the species can adapt and survive.
   This phenomenon does not imply, however, that individuals
of another species cannot survive. (Figure 16 in Chapter 18
illustrates the behavior of several species of fish under
different temperature conditions.) Thus we are now dealing
with several populations. Again we can arbitrarily divide
the fish into three classes: (1) sensitive, (2) intermediate, and (3) tolerant. The important distinction now
is that several species may be represented in each class.
There are some species (e.g., trout) where most individuals

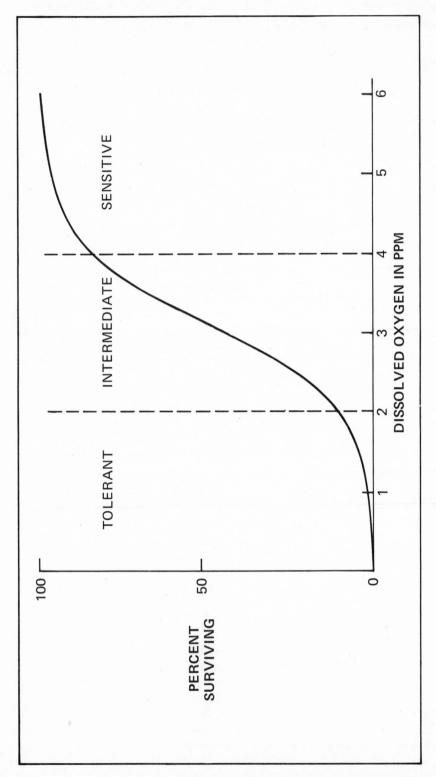

FIGURE 10.   *Diagram on the Survival of Aquatic Organisms under Various Concentrations of Dissolved Oxygen.*

would be in the sensitive group. On the other hand there
are species (e.g., carp) where most of the individuals will
be in the tolerant group. Thus increases in ambient
temperature will tend to select certain species for survival. It is important to note, however, that there is no
specific temperature where all individuals of species A will
survive and all individuals of species B will succumb.
This is another way of saying that the stress-effect curves
of various species will overlap.

Nature, however, does not impose changes limited to individual parameters of the physical environment. Whereas
individual parameters such as oxygen or temperature can be
varied and studied in the laboratory, under actual field
conditions many parameters will vary over any period of
time. Thus the surviving organisms are an indication of the
integrated adaptive capacities of these organisms in the
face of all prevailing environmental conditions.

BIOTIC FACTORS IN THE ENVIRONMENT

In addition to the various abiotic factors, organisms
are also exposed to numerous biotic factors in the environment. Perhaps foremost among these for primary consumers
is that of the food supply provided by the plants of the
system. The quantity and quality of the vegetation is the
key biological factor in the maintenance of the energy requirements of the total ecosystem under question. Thus any
physical factors which stress the plant community will have
a secondary biotic influence on primary consumers and
higher order consumers in that system.

A fundamental phenomenon in biology is that of *competition*. Plants compete for sunlight; algae compete for
nutrients; birds compete for territory; male deer compete
for female deer; carnivores compete for prey species. This
is a constant, dynamic process. If the system is changed
or displaced, there is a response, and a tendency to come
to a new equilibrium. Thus if overstory vegetation is re-
moved by an abiotic factor (wind) the understory will respond to the increase in solar radiation with increased
growth. The birds with the best ability to defend their
territory will be more competitive in establishing their
nests and rearing their young. Loss of vegetation through
plant disease will result in increased competition for habitat with a now-reduced carrying capacity.

Another important biotic factor is that of *predator-prey* relationships, previously noted in Chapter 4 under ecosystem concepts. It is clear that the predator is greatly influenced by the condition of the prey species. Prey species stressed by abiotic factors may become easy victims to predators. Conversely, predators in weak condition will often decrease their chances of survival. Predators with alternative sources of prey may have a better survival potential than predators with specialized prey relationships.

What we begin to see is a dynamic coupling of abiotic and biotic factors. Anything that disrupts the tendency to come to equilibrium will cause new responses in the system. This is one of the properties of ecosystems. Environmental stress, then, can be viewed as any physical event which tends to disrupt the established equilibrium of an ecosystem.

HUMAN-CAUSED FACTORS IN THE ENVIRONMENT

A logical extension of these discussions is the realization that humans may also influence the environment. Their various activities (see Chapter 5) can have significant influence on both the abiotic and biotic components of the environment. Large scale endeavors such as lumbering, agriculture, and irrigation cause major stress to the environment. These endeavors, of course, have been going on for hundreds of years.

In recent years humans have also been adding other factors to the environment, perhaps not quite as obvious. These include various chemical substances such as pesticides and heavy metals. This subject has become so important to environmental affairs that the next chapter is devoted entirely to environmental toxicology.

Looking at stress as we have in this chapter, it becomes apparent that environmental impacts are really a form of environmental stress. Recognition of abiotic and biotic factors in the total environment and how they may impose stress on an ecosystem is the first step in making an objective environmental assessment. Once the dynamic interplay of the existing environment is understood, it becomes easier to assess the potential impacts of human activities. This is the conceptual approach that is used in Chapter 12 on impact assessment.

Chapter 11

Environmental Toxicology

The decade of the 1970s was characterized to a great
extent by public concern for environmental quality. Major
environmental legislation was passed at all levels of
government during this period (see Chapter 2). Humans are
in a state of constant, dynamic interaction with the environment. Consequently, they affect the environment, and in turn
the environment affects them. Much of the current concern
focuses on air, water and land. Drinking water purity, ambient air quality, the use of pesticides, solid waste
disposal, dredge and disposal operations, oil spills, highway runoff, and many other issues are of increasing concern. Not only are we concerned about natural ecosystems,
including fish and wildlife, but today the health of humans
is also a major issue.

Because of the emotion that is frequently associated
with potentially toxic substances, it is often difficult to
assess the real hazards. This chapter attempts to outline
some important technical issues involved in the assessment
of such substances. Environmental toxicology is a complex
discipline which deals with the analysis and assessment of
potential harm from a variety of chemicals in the environment. The focus of this chapter will be on how certain
chemicals may be cycled through the environment, and which
properties of these chemicals are important in the assessment of potential toxic effects.

APPLICATION OF ECOSYSTEM APPROACH

Toxicology is the science of the noxious effects of chemical substances (see Appendix 2). Traditionally, the science of toxicology has been centered on the toxic effects of chemicals on humans and domestic animals. Today we recognize the broader scope of toxicology, including the newly-developing area of environmental toxicology. Chemicals move from air to water to land to vegetation, etc. These dynamic processes are more readily understood when they are examined in the context of cycling in ecosystems. Indeed, it was the application of the ecosystem approach that permitted an appreciation of complex processes underlying the cycling of potentially toxic chemicals in the biosphere. Therefore, the various concepts and processes in aquatic and terrestrial ecosystems discussed in earlier chapters provide a background for this chapter. Figure 11 is a simple diagram which illustrates how a potentially toxic substance may be cycled among various compartments of the environment.

TOXIC SUBSTANCES

A *toxic substance* is a chemical substance which is capable of producing death or serious harm to living organisms when present in relatively small quantities. Potential sources of these substances include air, water and food. In some cases toxic substances may be useful to humans (e.g., drugs, chlorine, pesticides); in other cases potentially harmful substances are introduced unintentionally into the environment (e.g., heavy metals, petrochemicals).

CHEMICAL NATURE

One of the primary, technical considerations in the assessment of potentially toxic effects from a chemical substance is the identification of the substances. This is often a difficult problem. Sometimes it is very costly and time-consuming to obtain accurate information. However, some attempt must be made to classify the substance, even if it is only classified in a general category (e.g.,

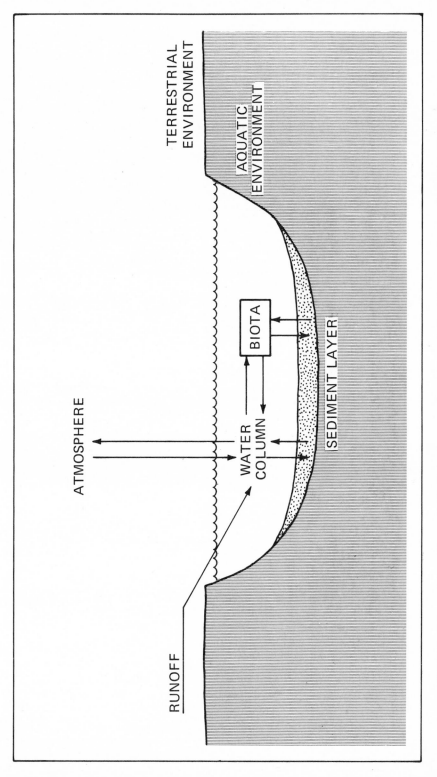

FIGURE 11.   *Diagram of Multi-Compartment Model of Movement of DDT Molecules in Various Ecosystems.*

mercury compound, benzene derivative). This is important
because the accurate assessment of any toxic effects depends
upon detailed information on the chemical properties of the
substance. This is discussed in more detail in subsequent
sections.

FATE OF CHEMICALS IN THE ENVIRONMENT

Where does a chemical go once it enters any complex
ecosystem? We are only now beginning to really appreciate
the complexity of this question. The best approach to
this complex problem is an ecosystems approach as discussed
above. It is usually helpful to picture any complex ecosystem as a *multi-compartment system*. Chemicals go into
various compartments, and rarely reach equilibrium. Consider the case of DDT, one of the more thoroughly studied
chemicals with widespread distribution in the environment
(Figure 11). Note how the DDT molecules can move from
compartment to compartment. The preferred pathways and
quantitative rates of transfer *(kinetics)* depend upon many
physical, chemical and biological factors.
    The fate of chemicals is complicated by the fact that
they also undergo *transformations* and *degradations*. For
example, some chemicals undergo transformations from one
chemical species to another. Mercury and polychlorinated
biphenyls (PCBs) are good examples. Other chemicals may
undergo degradation by some process; the photodegradation
of some pesticides is an example. Each new chemical species
will then behave differently in the ecosystem.

DYNAMIC CHANGES

Another complicating factor is the dynamic nature (or
the time dimension) of almost any chemical in the environment. The amount of substance changes constantly, especially as it goes from one compartment to another. DDT
again provides us with a good example of this issue of the
time dimension. DDT can remain in the mud or sediment
(organic phase) for decades; hence it is said to have a
long *persistence*. On the other hand, its physical properties are such that any appreciable amount in the water
column is only short lived (perhaps hours to days).

Soluble substances like some inorganic compounds may remain
in the water column for a long time. Toxic substances in
air from point sources undergo drastic dynamic changes in
concentration because of dispersion patterns. However, the
integrated effects of many sources in an urban environment
may maintain levels of toxic substances over relatively
long periods of time. As a rule, however, chemicals in the
environment experience *dynamic changes* more often than they
achieve a steady state.

CONCENTRATION

Another important concept in toxicology is that of *concentration* (amount/unit volume). When we know how much of
a substance is in the environment, we can make an objective
assessment of its probable effects. Allowing for the complicating technical factors outlined above, it is still
necessary to know the concentrations involved. Sometimes
the best estimate is the only information available, because
it is unlikely that we can know all the concentrations of a
given chemical in all the pertinent compartments of an ecosystem.

TOXIC EFFECTS ON ORGANISMS

If a chemical is used intentionally to kill or control
certain biological species, these species are called *target
species*. Other organisms in the area (or even in remote
areas) which may also suffer toxic effects are called *non-
target species*. What happens to any species will depend
upon:

1. The *chemical nature* of the chemical to
   which it is exposed.
2. The *concentration* of the chemical.
3. The *exposure time*.
4. The *toxicity* of the chemical for that
   particular species.

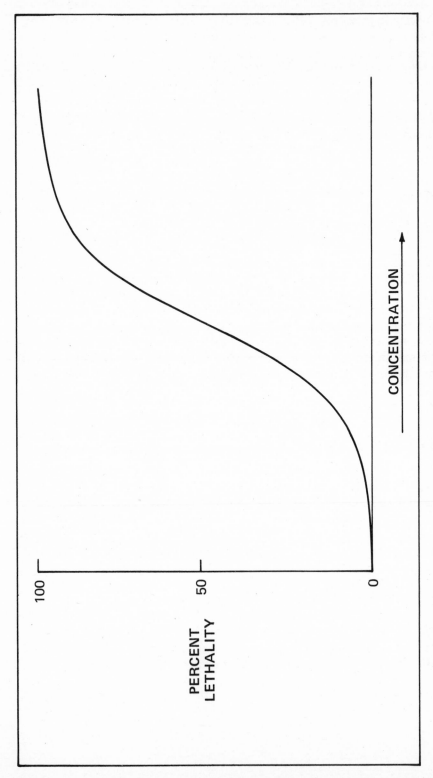

FIGURE 12.   Concentration – Effect Curve in which the Lethality Increases Directly with the Concentration of the Toxicant.

Knowing certain facts, then, it is possible to predict the probable effects. For example, knowing what the concentration of copper ion will be for a period of time, it is possible to predict what will happen to certain species in a pond which are exposed to it. Thus we speak of the *acute* (short-term) *toxicity* of a substance. Most chemicals will have a classical *concentration-effect curve* (see Figure 12). The acute toxicity of many chemicals is known, both for humans and for some animals, including some fish and wildlife. However, there are many chemicals for which toxicity data are not available or are available for only a few species.

More difficult is the assessment of the *subacute toxicity* and the *chronic* (long-term) *toxicity* of a given substance in the environment. Some substances undergo *accumulation* in a single organism *(biomagnification)*, or progressive accumulation in a food chain *(trophic magnification)*. Closely linked to the issue of accumulation is that of *metabolism*. Some substances are easily metabolized (chemically altered) and are excreted by organisms. Other substances may accumulate by build up of small increments, even though only trace amounts are in the environment. Chronic exposure and/or accumulation may then result in harm. Effects such as *mutagenic* (genetic-damage-causing) or *carcinogenic* (cancer-causing effects are often associated with toxic substances. Some compounds are *teratogenic* (deformation-causing) in nature. One of the difficulties of assessing chronic toxicity is that an end point such as death is not always apparent. Often a more common effect is a long term incapacitation, which may or may not lead to death. The understanding of the chronic effects of various chemicals in the environment is one of the biggest challenges facing the technical community today.

CLASSES OF TOXIC SUBSTANCES

There are literally thousands of potentially toxic substances, and no single compilation of these substances and their environmental effect has ever been prepared. The discussion below includes large, general classes of compounds that frequently receive close assessment in environmental studies.

1. *Organic Compounds*. This is a very large group of
    chemicals. Among the organic chemicals, various
    pesticides are of great importance to environmental
    studies. Pesticides include *herbicides* of various
    types. Herbicides are generally toxic to plants
    but not to animals. *Chlorinated hydrocarbon insec-
    ticides* such as DDT and heptachlor are poorly solu-
    ble in water, are generally stable, have long
    persistence in the environment, and thus accumulate
    in food chains. These compounds are toxic to
    insects and many aquatic organisms; they are con-
    siderably less toxic to humans. Another class of
    pesticides includes the *organophosphate insecti-
    cides*. These are soluble in water and generally
    unstable. However, they tend to be highly toxic
    to both aquatic and warm-blooded species. Most
    insecticides exert their toxic effects on the
    nervous system of animals. Closely related to
    pesticides are the *polychlorinated biphenyls* (PCBs);
    they have many properties similar to those of DDT.

2. *Petroleum Hydrocarbons*. Petroleum hydrocarbons are
    also organic compounds, but because of general in-
    terest they are often treated separately. Petroleum
    hydrocarbons are complex mixtures of *aliphatic*
    (straight-chain) and *aromatic* (cyclic) molecules.
    They have high vapor pressure, are poorly soluble
    in water, and usually have low density. Toxicity
    varies widely, depending upon the species of
    animal and molecular composition. Many of the
    aromatic hydrocarbons found in petroleum are highly
    toxic, and some are thought to be carcinogenic.

3. *Inorganic Chemicals*. This is another large group
    of compounds, and it contains many acids, bases, and
    salts, most of which are highly soluble in water.
    Consequently, they tend to be common in surface
    waters. *Ammonia* is highly soluble and is a product
    of the decomposition of organic compounds. It is
    generally toxic to aquatic organisms. Chlorine
    is a gas which is soluble in water. It also has a
    high vapor pressure, and thus tends to evaporate
    into the atmosphere. Chlorine is toxic to aquatic
    species, and is commonly used as a biocide. Other
    substances which may be toxic at low concentrations
    include *cyanide, nitrate,* and *nitrite* in water. A
    special case is that of *sulfur dioxide*. Burning
    fossil fuels may emit large quantities of sulfur

dioxide, which combines readily with water moisture to form acids. These various acids (e.g., *sulfuric acid*) are toxic to plant tissues and to the lungs of warm-blooded animals.

4. *Metals*. Metals and their various salts are also generally soluble in water. Trace amounts of certain metals (e.g., *cobalt, copper, zinc*) are essential for life because they form important components of organic molecules such as vitamins and enzymes. However, in higher concentrations certain heavy metals and some essential metals are toxic. Toxicity may include both acute and chronic effects. Heavy metals such as *cadmium, copper, lead, mercury* and *selenium* have received special attention in recent years. These metals may be toxic to humans, causing damage to liver, kidneys, and the nervous system.

IMPACT ASSESSMENT

With the recent enactment of the *Toxic Substances Control Act* (see Chapter 2) additional emphasis will be placed on toxic substances. There is good evidence that the engineering profession is now paying increased attention to the molecular hazards from various projects. In the earlier years after the enactment of NEPA and other laws, there was a learning period for the technical community; most of the attention was focused on macrobiology and the more obvious impacts. Now environmental assessments are beginning to include various toxic chemicals which can cause damage to the biosphere.

A fundamental concept in toxicology that deserves stressing is the importance of the concentration or amount of chemical. Mere presence of a toxic substance does not mean that a real hazard exists. For example, traces of heavy metals in surface waters are practically universal. Traces of DDT in human body fat are the rule rather than the exception. It is the concentration and the amount of chemical that will determine the toxic effects. Even pure water, if taken in excess quantity, can be acutely fatal to humans. Thus the degree of potential hazard or harm is very much a function of the quantity of the chemical that

will reach the non-target species. Much of the public
misunderstanding of the real hazards associated with toxic
substances would be avoided if this concept were communicated in an effective manner.

On the other hand, there may be real hazards associated
with long term exposure to very low levels of certain chemicals. There is accumulating evidence that low levels of
certain chemicals can cause stress to the biosphere,
exerting influence on various cellular, physiological, and
ecological processes. For example, some human cancers
and various abnormalities in aquatic animals are thought to
be caused by chronic exposure to chemicals in the environment. More information is vitally needed before we can
evaluate such hazards in a responsible manner in the years
to come.

Chapter 12

Impact Assessment

There is no single guideline which has received universal acceptance for conducting an environmental impact assessment. Individual agencies and numerous academic and consulting groups have developed various guidelines and methodologies. However, no single approach has been adopted as offering superior advantages over other methods. Consequently, impact assessment is still an art requiring much judgement and experience. This is true both for the assessment of impacts on biological systems, and for the more comprehensive efforts required for the assessment of impacts on the total environment.

SOME COMMENTS ON IMPACT ASSESSMENT

It is a remarkable fact that many years after the enactment of NEPA and other environmental legislation, the environmental research community still has not achieved a high standard for environmental assessment. One of the major reasons is that studies are often disjointed and poorly coordinated. Some components are studied with great precision and sophistication; others are glossed over in an inadequate manner. Commonly the individual components are not related in an overall, coordinated assessment. Biologists have contributed their efforts to poorly-executed studies. A common failure in the past has been an emphasis on the compilation of inventories of flora and fauna without assessment of their relevance to the proposed project. The failure has not been in the degree of professionalism applied to the study of individual groups of biota. Rather the common failure has been in the lack of systematic team

efforts to relate the biological data to all the other
environmental components. This same failure is also common in the study of the other environmental components (e.g.,
geology, air quality, water quality). In summary, the
strength of environmental studies lies in technical *analyses*
within individual disciplines and specialties (e.g., fisheries biology, air dispersion analysis, thermal analysis); the
weakness lies in the *integration* of multidisciplinary and
interdisciplinary efforts. Not only is such integration
good procedure, it is also mandated in Section 102 of NEPA
(see Chapter 2), where it is stated that a *systematic,
interdisciplinary approach* shall be utilized.

THE SYSTEMS APPROACH TO IMPACT ASSESSMENT

One solution to this common problem is to adopt and
apply a *systems approach* to environmental studies. An important feature of the systems approach is that it emphasizes *system optimization* rather than *component optimization*.
Also, the systems approach offers a conceptual framework
for the planning, development, and coordination of an
environmental study. A sound reason exists for each input
element, and a conceptual framework for its integration is
provided.

Figure 13 is a diagram of a systems approach to an
impact assessment process. Note the three critical inputs:

    various agency guidelines
    data on the existing environment
    a description of the proposed project and its
        alternatives

The impact assessment process usually results in prelimin-
ary assessments, often with a feedback for additional
environmental data and project description. Further,
refined assessments result in the final impact assessment.

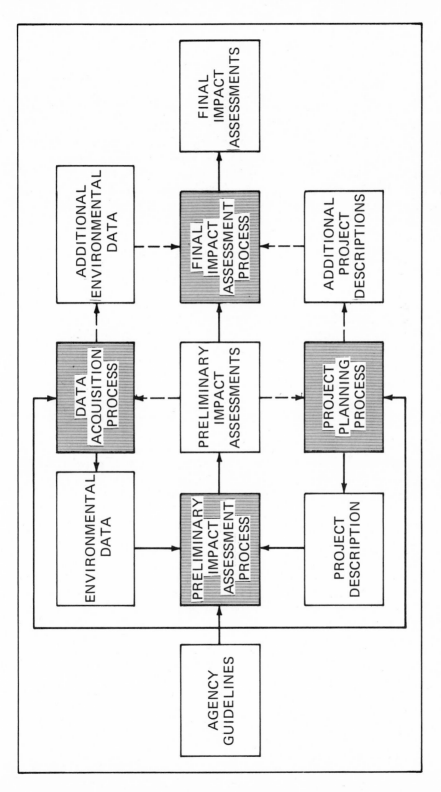

FIGURE 13. *Diagram of a Systems Approach to an Impact Assessment Process.*

ABIOTIC AND BIOTIC IMPACTS

The following discussions illustrate the assessment of the potential environmental impacts that would result from a proposed project and its alternatives. The need to predict such impacts with the available state-of-the-art is a requirement under NEPA and other legislation. Engineering projects will generally result in localized changes to the air, the hydrology, and the land mass. Hence they will cause various *abiotic impacts*. Because engineering specifications of the proposed actions can be developed, it is possible to predict the abiotic impacts on the existing physical environment with some degree of accuracy.

It is also possible to predict some of the *biotic impacts* with a fair degree of accuracy. Many biological organisms are non-motile (e.g., trees, periphyton) or relatively non-motile (e.g., benthic organisms), and hence the direct impacts of physical changes can be predicted. Thus, for examples, land clearing, rechannelization, and draining will have direct impacts on the biota in the immediate project area. Much more difficult is the prediction of the probable impacts on biota that are mobile (e.g., birds, mammals, fish). Often these mobile species are the very species of interest to fish and wildlife agencies and to the public (see Chapter 9). However, based upon knowledge of habitat requirements it is possible to make some reasonable predictions of the probable behavior of mobile populations.

IMPACTS ON HABITAT

As discussed in Chapter 5, habitat is the physical location where biota actually exist. Thus the direct biotic impact on vegetation (producers) will also have an indirect effect on mobile consumers that require the vegetation for habitat. Heavy siltation of a stream may have both direct and indirect impacts on fish that utilize the stream as habitat. Predictions of such impacts are difficult because there are many interactions among the abiotic and biotic elements in various habitats. However, based upon ecosystem concepts and an evaluation of the probable carrying capacity (see Chapter 9) of the habitat, it is possible to make predictions on the probable impacts on the biota supported by a given habitat. Also, the existence of any critical habitat for relatively rare species and/or species officially listed in the threatened and endangered categories can also be evaluated. It is important to point out that habitat assessment is a difficult and specialized process, and disagreements among professional biologists are common.

ECOSYSTEM IMPACTS

Any distinctions among biotic impacts, impacts on habitat, and ecosystem impacts are necessarily vague. However, if we go back to our earlier ecosystem concepts (Chapter 4), we may anticipate that the evaluation of ecosystem impacts requires the assessment of the dynamic interactions among the various abiotic and biotic elements of the system. Predictions of probable ecosystem impacts are more difficult than those centering on biotic components. However, there is increasing demand to assess ecosystem impacts, because such assessments are valuable in predicting probable, longer term trends in the ecosystem. The types of ecosystem attributes that may be assessed include the following:

productivity	density
food chains/food webs	diversity
community structure	stability

The state-of-the-art in making predictions of impacts on the regional biosphere is progressing rapidly. Consequently it is important to extend predictions of impacts beyond the direct impacts immediately obvious. Figure 14 is a simple diagram of how the proposed project description and ecosystem theory are combined to make the various impact predictions.

IMPACT "DIMENSIONS"

The impact assessment process has matured sufficiently to enable us to be more definitive in our impact assessments than we were only a few years ago. It is no longer adequate to say an impact will occur; it is necessary to predict certain *impact "dimensions"*.

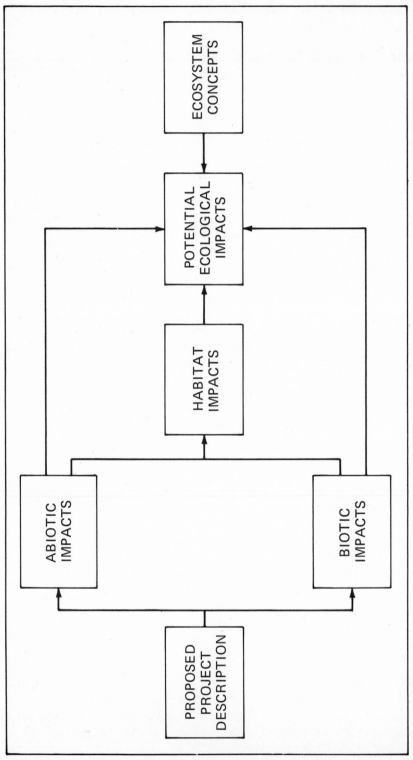

FIGURE 14. *Diagram of How the Proposed Project Description and Ecosystem Concepts May be Applied to the Prediction of Potential Ecological Impacts.*

1. *Probability.* It is generally possible to predict
    the probability that an impact will occur. This
    does not mean that probability theory can be
    applied to all impact predictions, because clearly
    this is not the case. However, as discussed above,
    the probability of some biotic impacts in the
    project area is virtually certain. Subsequent
    impacts are more difficult to predict, but the
    frequent need to predict the probability of
    specific impacts requires our best efforts.
2. *Time Frame.* Another impact dimension is that of
    the time frame of the impact. It is clearly im-
    portant to assess whether an impact is *very short
    term* (e.g., days to weeks), *short term* (e.g., 1-3
    years) or *long term* (e.g., several decades). De-
    cisions on projects are more meaningful if the
    time frame of potential impacts is assessed.
3. *Magnitude.* Finally, it is critical for environ-
    mental specialists to predict the probable magni-
    tude of the impacts. Some impacts can be quanti-
    tated readily, such as affecting a specific number
    of acres of forest or wetland. Others, such as the
    percent of a deer herd that may be lost, are more
    difficult to quantitate. Acres of exposed construc-
    tion site can be quantitated; potential increases in
    turbidity values in water are more difficult to
    quantitate. Nevertheless, the need to quantitate
    is now mandated by the improving state-of-the-art in
    making impact predictions.

MITIGATION MEASURES

The concept of *mitigation* of environmental impacts is
firmly established in environmental legislation and agency
guidelines. Mitigation is not an ecological concept, although mitigation techniques are often based on sound
ecological theory. Mitigations of potential ecological impacts will frequently center on impacts that are considered
significant *adverse impacts*. Thus they are relevant to the
specific region. Some agencies (e.g., U.S. Fish and Wildlife
Service) have some general policies on mitigation measures.
But for the most part, mitigation measures for potential
ecological impacts are quite project-specific. There is
considerable latitude for resourceful and innovative

mitigation approaches, in addition to the application of
well-established measures. This subject is of great importance to both engineers and biologists, and a whole
chapter (Chapter 19) is devoted to mitigation measures.

APPLICATION OF IMPACT ASSESSMENT TO VARIOUS PROJECTS

   Subsequent chapters of this book examine potential
impacts associated with various engineering projects. The
emphasis is on examining the potential impacts on the regional biota resulting from various construction activities
and operation and maintenance activities. Since many effects on the abiotic environment (e.g., surficial geology,
hydrology) have critical importance to the regional biota,
it is incorrect to describe such effects as being outside
the realm of biology. Indeed, emphasis on the ecosystem
approach minimizes the risks of looking only at impacts
upon wildlife and their habitat. Consequently, comments
on the potential impacts associated with these various projects include a wide spectrum of abiotic, biotic, habitat,
and ecosystem effects . Finally, some references are made
to important legislation, pertinent guidelines, and technical
documents of relevance to the application of biology to
engineering projects.

Chapter 13

The Total Human Environment

One of the common consequences of complete immersion in the details of environmental studies is the loss of perspective on the total human environment. As fish surveys are being conducted, as estimates of deer herds are being developed, and as studies on the productivity of wetlands are underway, we often lose sight of the big picture. Consequently, it is valuable to review some important factors which are highly relevant to the total efforts of any project.

LEGISLATION AND REGULATIONS

Perhaps the strongest message on the need to consider the total human environment can be found in the language of Section 102 of the *National Environmental Policy Act of 1969* (see Chapter 2). The language is unambiguous, and it sets a policy tone which requires a total overview of the consequences of any proposed action. It does not single out any environmental components (e.g., fish and wildlife) for preferential scrutiny. On the other hand, as has been demonstrated by court actions, it does not allow for an inadequate assessment of any environmental components.

Another important example can be seen in the language of the *Federal Water Pollution Control Act Amendments of 1972*. This important Act has had profound impact on the total environmental affairs community of the United States. Industry, professional engineers, professional biologists, advocate groups and public agencies have all added their viewpoints in the implementation of the mandates of the Act. There have been times when the total objectives of the Act

have gone out of focus. Again, the language of the Act
(e.g., Section 304) makes it clear that human objectives,
such as health, welfare and aesthetics, are included within
the provisions of the Act.

More recently, the President of the United States has
issued *Executive Order 11990* (May 24, 1977) in relation to
the protection of the wetlands of the Nation. This Order
clearly outlines a National objective to minimize the loss
and destruction of wetlands. However, the Order also
directs that factors of relevance to the human condition
shall also be considered. Thus there is an attempt to
provide a balanced assessment which examines both the
physical environment and the social environment of humans.

These examples are presented as an overview of the
intent of much of the environmental legislation. At different times and in different regions there have been
preferential emphases on certain environmental components.
The important point, however, is that the technical efforts
of environmental biology must be integrated into the overall efforts, and these efforts generally examine the total
human environment.

COMMON ENVIRONMENTAL COMPONENTS

Biologists working with other team members will
generally study a number of abiotic and biotic components
of the regional environment. Details will depend upon
agency guidelines, the regional ecology, and specific project requirements. However, the following are among the
environmental components which are generally included in
a study:

bedrock geology	terrestrial habitat
surficial geology	aquatic habitat
water quantity	wetland habitat
water quality	air quality
aquatic flora	precipitation
aquatic fauna	temperature
terrestrial flora	wind conditions
terrestrial fauna	

These components describe the *physical environment*. In
addition, various other components are generally assessed
in the *social environment*. These may include:

    demography
    economics (including agriculture)
    transportation
    utilities
    sociology
    public health and safety
    educational resources
    historical resources
    archeology
    cultural resources
    recreational resources
    conservation
    preservation
    aesthetics

The physical environment and the social environment are
components of the *total human environment*. Many environmental projects, especially those conducted under NEPA, are
required to assess the total environment. Consequently,
the project team must understand that the biological studies
are only one part of the overall assessment effort.

WHAT ARE "SIGNIFICANT IMPACTS"?

Two recurrent realities face the professionals involved in environmental assessments. In the first place,
a proposed project and its various alternatives may cause
literally dozens of potential impacts. This is true for
both biological impacts and other types of impacts. Another
reality is the frequent requests biologists receive to
single out the "significant impacts". How does one do
this? There is no single formula, and the answer is influenced by both practical and philosophical considerations.
A practical approach rests to a great extent on the
assessment of the *probability, time frame,* and *magnitude* of
the potential impacts, as outlined previously in Chapter 12.
Some sort of integration of these various impact dimensions
is required, and a number of approaches have been proposed.
However, it is important to note that a "quantitative"
assessment based on numerical "values" for many impact

TABLE 5. *Example of Summary Matrix Used to Summarize Impacts on the Physical Environment in Relation to the Human Environment.*

Selected components of physical environment that will experience impacts	Selected components of the human environment							
	Regional economics	Land values	Social institutions	Public health	Educational resources	Recreation	Conservation	Aesthetics
Air quality								
Noise								
Hydrology								
Water quality								
Surficial geology								
Aquatic biota								
Terrestrial biota								
Wildlife habitat								

parameters is really semi-quantitative and often subjective. Much opposition has been generated to this approach. Nevertheless, it is possible to apply common sense and to use these impact dimensions for an estimate of the more important impacts.

Another practical approach is to consider the various *regional environmental factors*. Loss of 50 acres of wildlife carrying capacity on agricultural land in a region of thousands of acres of comparable agricultural land may be insignificant to the regional ecosystem. On the other hand, loss of one acre of wetland in a region with strong conservation policies may be highly significant. Thus, again, common sense and a knowledge of the local human environment will help to dictate what is potentially "significant". Table 5 shows a simple summary matrix used to summarize various physical impacts in relation to elements in the human environment. These environmental elements are frequently of significant interest to people, although the details vary on a regional basis.

THE PROFESSIONAL SCOPE OF BIOLOGY

An important philosophical issue emerges from the foregoing discussions. The field of biology applied as a physical science cannot give opinions on "good" or "bad". As a physical science it deals in natural phenomena and facts analyzed through the application of biology, chemistry, physics and mathematics. However, the legislation upon which most environmental studies are based requires that the studies be relevant to the human condition. Thus biological and ecological impacts must eventually be assessed for their significance to the total human environment, as discussed above. It is here that biologists and ecologists must integrate their efforts with the interdisciplinary team to assess both positive and adverse impacts in relation to the local, human environment.

A biologist is not trained to assess socio-economic factors, agriculture, public health, and recreational alternatives. These are all factors that often interact with the regional ecology. Persons trained primarily in the biological sciences cannot be expected to give valid, professional assessments of environmental components outside their expertise. Such assessments are the responsibilities of the entire environmental team. This is why coordination is so vital.

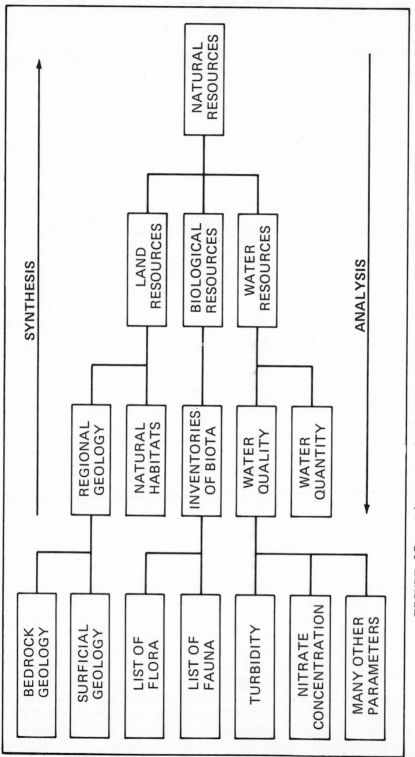

FIGURE 15.   *Diagram of the Reciprocal Relationship Between Analysis and Synthesis in the Environmental Assessment Process.*

The practical pressures of conducting environmental studies frequently impose responsibilities for the assessment of several environmental components on individuals without extensive training and experience in all such components. For example, it is not unusual for a "biologist" to compile and interpret data on meteorology, hydrology, water quality, fisheries, geology, terrestrial vegetation, and wildlife. This procedure is not necessarily wrong, because the limited scope of many projects will dictate a common sense approach in the selection and size of the project team. The limitations of this procedure become more significant as the magnitude of the project increases. The project team must have a clear understanding of the professional scope of biology, and which individual specialties (see Appendix 2) are required to conduct an acceptable study for a given project.

ANALYSIS VERSUS SYNTHESIS

Under the general comments on impact assessment in Chapter 12, there was a passing reference to the tendency of most environmental studies to stress *analysis*. And yet the more relevant the study becomes to the total human environment, the greater is the need for *synthesis*, or the integration of the data from various environmental components. This reciprocal relationship between analysis and synthesis is illustrated in a conceptual diagram in Figure 15.

The conventional approach of a technically trained team is to assess the natural resources of a region through analysis. For example, the water resources are analyzed for the number of acre-feet of impounded water and perhaps the number of miles of stream channel of flowing water. The water resources are also analyzed for quality; perhaps several dozen parameters (e.g., turbidity, nitrate) would be analyzed. The biological resources are routinely analyzed through inventories of biota, down to lists of individual species. Land resources, in turn, may be analyzed through such analytical components as open land, forested land, agricultural land, surficial geology and bedrock geology. The analysis can obviously go into great detail.

Citizens, agency heads, and others involved in the
review and decision-making process are interested in the
reverse process. For example, a state governor, a regional
administrator, or a corporate official may want to know
what an existing or a proposed facility will do to the
natural resources, the human resources, and the economics of
the region. Any interest in such details as turbidity,
bedrock geology, and herbaceous vegetation becomes subordi-
nated to the need to see the total picture in perspective.
Recent regulations (e.g., the CEQ, 1978; see Selected
References) are stressing the need for more integrative
assessment. This synthesis of information is the most difficult task faced by the technical team. Note in Figure 15
that a summary overview of natural resources requires (but
is not limited to) an integration of habitat, water quality,
lists of biota, and the regional geology. The professional
biologist is generally not trained to do this integration.
However, other scientists such as geologists, agronomists,
hydrologists and agricultural economists may also be in-
experienced in this synthesis. For that reason, careful
organization of the research team is required to produce
a well-thought-out and balanced assessment of the total
human environment. This is one of the most challenging
facets of the entire process of the application of environmental studies to decision-making.

How the study team is organized to achieve the synthesis
function is well beyond the scope of this book. What is
important is that project managers recognize and be alert
to this common limitation in the professional backgrounds
of most scientists and engineers. Additional comments on
this important subject appear in the general guidelines
outlined in Chapter 20.

Chapter 14

Transportation Systems

In a modern society transportation systems encompass a vast network of highways, airports, railroads, waterways and port facilities. In recent decades, the rapid development of automotive and air transportation has resulted in rapid expansion of the highway system and airport facilities in the United States. Much of this expansion is still going on, although at a slower rate. Rail facilities are actually being reduced in many areas of the country. However, the current need for urban mass transit systems has resulted in renewed interest in existing railroad beds. Water transport continues to be an important element in our total system. Most of our current engineering activities, however, are related to maintenance of waterways, rather than the construction of new waterways. Because of the relative emphasis on highway development, most of the discussions in this chapter will center on highway projects.

HIGHWAYS AND BRIDGES

Most of the environmental assessments for proposed Federal actions under NEPA (see Chapter 2) have been carried out for highways and bridges. More environmental impact statements have been prepared for highway projects than for any other type of project. The U.S. Department of Transportation, Federal Highway Administration has prepared many guidelines, manuals, and training courses on various facets of highway impact assessment. Included among these has been a workbook and training program entitled *Ecological Impacts of Proposed Highway Improvements*. About 1200 Federal and state highway professionals have attended this course. In addition, the Federal Highway Administration has sponsored the preparation of a practical handbook entitled *Highways and Ecology: Impact Assessment and Mitigation* (Erickson *et al.*, 1978). Thus there has been considerable commitment to train highway professionals in ecological concepts and their practical applications to highway projects.

Highway improvement projects in highly urbanized areas generally do not require extensive assessments of impacts on biota. This is especially true if the highways stay within the existing rights-of-way. More extensive improvements, especially with several, proposed, alternate alignments, may require some attention to biological resources. But as a general rule, the potential impacts on biological resources, especially wildlife, are relatively minor in the urban setting.

As the scene shifts to the suburbs and to rural areas, the assessment of potential ecological impacts becomes much more important. Projects frequently include major interstate highways and state highways. There is need to consider alternate corridors and even alternate alignments within general corridors. Major changes in the terrain and in vegetation become highly probable, especially for multi-lane highways of considerable length. Also the proposed alignments frequently traverse a variety of habitats such as stream habitats, open fields, and forested areas. Thus there is need to examine the short-term impacts from the construction phase and the longer-term impacts from the operation and maintenance phase.

CONSTRUCTION PHASE

The construction phase of a highway will typically extend over a period of several years. During this period a number of construction activities may take place, including the following:

    land clearing (removing vegetation, chipping)
    stripping
    drilling and blasting
    excavation
    filling
    draining
    dredging
    rechannelization

disposal of spoil
placement of foundations and footings
pile driving
pouring concrete
bitumen paving
materials processing (e.g., gravel)
trucking materials
mulching
seeding and planting
fertilizing

These are all physical activities and as such they can generate noise, dust, and erosion runoff; they can eliminate and/
or disrupt wildlife habitat; they can cause direct losses of
biota; and they can cause the introduction of potentially
harmful chemicals into the environment.

OPERATION AND MAINTENANCE PHASE

The normal operating life of a modern highway is several
decades, and consequently the operation and maintenance of
highways may be viewed as long-term activities. Certain
practices may have potential impacts associated with them.
Let us consider some of these customary practices:

repaving
mowing
slope stabilization
reseeding
snow removal
use of salts and sand
use of paints and preservatives
application of pesticides

In addition to these activities normally associated
with maintenance, the actual operation of a highway will
cause various abiotic and biotic effects, such as: changes
in air quality; changes in noise level; changes in water
quality from highway (non-point source) runoff; incidence
of accidental spillage; and roadkill of large mammals.

Not all of these effects will be significant to the
regional ecosystem, based on some of the concepts and criteria discussed in Chapter 13. As a general rule it is good
practice to identify and assess those impacts that are
likely to be significant to the regional biosystems. However, it is difficult to predict what will have relevance
to the local, human environment, and thus it is important to
be alert to the local environment as well as to general
principles and general guidelines.

POTENTIAL IMPACTS OF COMMON CONCERN

Based upon generalized experience with various regions
of the country, an attempt can be made to summarize some of
the potential impacts commonly associated with highway development projects (Table 6). These impacts have been
selected as being of predictable, recurrent concern to the
following:

    highway professionals
    regulatory agencies
    fish and wildlife agencies
    organized citizen groups
    individual citizens

Note that these impacts include abiotic impacts, biotic
impacts, impacts on habitat, and longer term ecosystem impacts.
Specific comments on various impacts listed in Table 6
based upon some individual cases of public concern appear
useful. Concern with potential disruption of breeding behavior resulted in the limitation of the seasons acceptable
for construction in one project. In another case a woman
at a public hearing announced that she knew that the highway
could not be built unless it conformed to air quality standards; what she wanted to know is why an assessment had not
been done on the potential impact of air quality on her
sensitive plants. At another public meeting a major concern
was with the potential impacts of road salt on public water
supply and hence on people with heart disease. Note that
good standard practices to seed and fertilize a construction
site will have positive, long term results, but may have
adverse, short term impacts on water quality. Other issues
such as potential impacts on wetlands, roadkill of mammals

TABLE 6. *Summary of Potential Short Term and Long Term Impacts Frequently Associated with Highway Projects.*

Project activities	Potential impacts	
	Short term	*Long term*
Land clearing and stripping	Loss of primary producers. Effects of erosion runoff on aquatic community.	Loss of carry capacity. Changes in density and diversity of wildlife.
Drilling, blasting, and excavation	Disruption of nesting and breeding behavior of birds and mammals.	
Cut and fill operations	Changes in hydrology, surficial geology, and vegetation.	Changes in productivity of aquatic habitat.
Rechannelization	Loss of aquatic biota. Effects on water quality.	Changes in productivity of aquatic habitat. Changes in diversity of aquatic community. Loss of aesthetic value.

TABLE 6.   (Continued)

Project activities	Potential impacts	
	Short term	Long term
Seeding, fertilizing, and slope stabilization	Effects on water quality.	Reduction in erosion; improvement of wildlife habitat.
Mowing within right-of-way		Effects on terrestrial habitat and species diversity.
Use of pesticides within right-of-way	Effects on water quality and aquatic community.	Effects on terrestrial habitat and species diversity. Effects on water quality and aquatic community.
Use of salts and sand	Effects on water quality and aquatic community.	Effects on human health. Effects on vegetation.

TABLE 6. (Continued)

Project activities	Potential impacts	
	Short term	Long term
Highway traffic	Effects from accidental spillage of chemicals.	Effects of air quality loss on sensitive plants. Effects of noise on animals. Effects of roadkill on populations of large mammals.
Existence and operation of highway		Potential changes to critical habitat and unique habitat for rare, threatened, and endangered species.

(especially deer), and the loss of critical habitat are highly probable, and much of our efforts for impact assessment
and mitigation are frequently centered on such environmental
issues.

Frequently an overlay of other regional factors such
as conservation, preservation, aesthetics, land use policies,
and economics must be integrated into the ecological impact
assessments (see Chapter 13). The need for the total team
approach becomes critical. The more controversial the project, the greater the necessity to coordinate the total team
efforts.

AIRPORT PROJECTS

Airport projects have also required extensive environmental assessments, but many of these projects have been ex-
pansions and/or improvements of existing facilities. Also
many of these projects have been in semi-urbanized areas.
Consequently some abiotic factors and certain socio-economic
components tend to receive more emphasis than biotic components. The effects of noise, air quality deterioration,
parking requirements, and access highways receive emphasis.

There are instances, however, when new airport facilities are proposed. Often such facilities are significant
distances from the metropolitan centers. Also such facilities tend to be quite large. These factors combine to make
the potential impact on biological resources highly significant. The clearing and paving of the 1000s of acres of land
will have important consequences to the regional ecosystems.
Thus extensive assessments of the abiotic and biotic environment are required.

Frequently the overlay of other regional (or even
national) factors such as conservation, preservation, aesthetics, and land use policy mentioned above will also enter
into the overall assessments of airports. The larger the
project, the greater the probability for conflicts. The
classic example is the large airport proposed in the Ever-
glades in Florida. This project received national attention,
and was eventually determined to be an undesirable alternative.

MITIGATION MEASURES

Mitigation measures for various projects are discussed in Chapter 19. However, it is useful to conclude this chapter by highlighting the types of mitigation measures frequently proposed for projects that may involve significant areas of land, such as highways and airports. Included are the following:

1. Minimize erosion and runoff through good design and construction management practices.
2. Minimize the loss of terrestrial habitat.
3. Minimize the loss of aquatic habitat.
4. Avoid wetlands.
5. Avoid critical or unique habitats.
6. Minimize the potential for roadkill.
7. Minimize the use of pesticides.
8. Minimize the use of road salts and sand.

Many of these measures have been in existence for decades, long before NEPA and other environmental legislation. For example, sedimentation basins and the use of culverts have been standard practices for many years. Today, however, there are excellent opportunities for innovative approaches to mitigate impacts on biota, and many, creative, joint efforts between engineers and biologists are now appearing nationally.

Chapter 15

Utility Corridors

Closely related to highways are various types of utility corridors. Utility corridors have been in operation for decades, and historically the main concerns have centered on land use, rights-of-way, property values, and other economic issues. In the past decade environmental issues have also entered into the picture. The trans-Alaska oil pipeline environmental assessment (see Chapter 3) raised the interest in the oil pipeline corridor to a national level. A new plan to transport gas from Alaska is being implemented currently, and several corridors were considered. Throughout the west and midwest, the new energy development projects are receiving close attention because of the need to carry electric power over long distances with transmission lines. Private utilities are now finding the need to consider numerous environmental issues, in addition to the more traditional concerns for economics and reliability, in their selection and maintenance of utility corridors.

TRANSMISSION LINES

It is estimated that about 300,000 miles of overhead electric transmission line corridors are controlled by electric utilities in the United States. The rights-of-way associated with these corridors comprise millions of acres of land. Thus there is extensive potential for wildlife habitat within these corridors. The U.S. Fish and Wildlife Service is currently investigating the potential to manage these corridors for fish and wildlife habitat. Consequently there is great interest in the biological resources within transmission line corridors.

The construction activities for transmission lines are
not as disruptive of the physical terrain as those associated
with multi-lane highways. The placement of poles and towers
requires some excavation, but the areal extent is relatively
small. Some clearing and grubbing takes place, but this
depends upon the terrain and the transmission system. The
hauling of construction materials and the establishment of
the construction camp generally may cause localized disturbance of flora and fauna. Some drainage ditches are frequently necessary. However, the amount of erosion is generally small compared to highway construction. Nevertheless,
short term disruption of aquatic and terrestrial wildlife is
common.

Some of the major concerns with ecological impacts of
transmission line corridors center on the management of the
corridor. Herbicides were used extensively in the past to
maintain a clear right-of-way. This practice resulted in
the loss of vegetation and hence carrying capacity. Also, it
is difficult to control the application of herbicides, and
consequently areas outside of the right-of-way were often
affected. On the other hand, the areas relatively clear of
overstory vegetation frequently have a good diversity of
shrub vegetation and other understory vegetation. This, in
turn, maintains a more diverse food web than the forest alone.
Thus, the cleared right-of-way maintains an ecotone and in-
troduces increased species diversity along the corridor.
Therefore the details of management of the utility corridor
will determine the potential impacts on the regional ecosystem. There is a tendency today not to use herbicides, at
least not to the same extent as formerly, and thus allow the
natural process of ecological succession (see Chapter 5) to
occur within the corridor.

Another area of potential biological concern is associated with the high tension of some electric transmission
lines. The increasing need for electrical energy in the
United States is resulting in electric power transmission
requirements with increasing voltages. Transmission lines
with voltages in the range of 750-1500 kV are now operational,
and offer certain benefits in costs per unit of power trans-
mitted. At the same time there are physical phenomena associated with these high-voltage systems including (1) noise,
(2) corona effects, (3) electric fields, and (4) magnetic
fields. At the relatively lower voltages (below 500 kV)
there probably is no significant effect on wildlife behavior.
However, the influence of the higher voltages on plants and
animals is not known. This is currently an area of research,

and it emphasizes the fact that our state-of-the-art in some technologies is often well ahead of our understanding of the total environmental implications of that technology.

COMMUNICATION LINES

Various communications facilities also use utility corridors. These include telephone lines and wireless communication systems such as VHF and UHF relay stations. Modern electric utility systems frequently have radio communication systems as part of their system of extensive communication and control. These radio communication lines frequently follow the same routes used for the transmission lines.

Construction of communication lines is generally not highly disruptive of the surficial geology, vegetation, and wildlife, although details depend upon the specific system. Wireless communication lines use microwave towers, and construction is generally localized. Clearing of access roads is frequently necessary. Telephone lines may require more extensive clearing, both for the initial construction and the subsequent maintenance of the lines.

Some of the same concerns with the management of right-of-way corridors for transmission lines also occur with communication lines. The use of herbicides for the control of vegetation along the right-of-way may have adverse effects on wildlife habitat. However, the potential for habitat diversity and understory vegetation for wildlife along the corridor should be recognized. Frequently microwave systems can be maintained in such a way that only the tops of trees that obstruct the line-of-sight between towers are cut. This results in minimal disturbance of the vegetation and thus the wildlife habitat.

OIL AND GAS PIPELINES

Oil and gas pipelines also constitute thousands of miles of corridors in the United States. Moreover, oil and gas pipelines have been in operation since the last century. Consequently there is considerable experience with the construction and maintenance of pipelines.

The construction of large pipelines can cause significant effects on the abiotic and biotic components of the
corridor. The relatively deep excavation will disrupt the
surficial geology, hydrology and vegetation. Drainage
ditches and access roads are also frequently necessary.
Long pipelines require extensive construction camps. Extensive noise from the construction equipment and from
blasting may cause as much disruption of wildlife as is
frequently experienced with highway construction.

One of the important aspects of the construction of long
pipelines from the biological viewpoint is the potential to
traverse many types of habitats. As with highways, the probability of open fields, agricultural land, forested areas,
streams and wetlands increases with the length of the corridor. Thus the complexity of the ecosystem analyses increases. Also the probability of specialized or even critical habitat increases. Threatened and endangered species
become a highly predictable concern with such projects.
Very long corridors may even traverse several biomes, and
thus a whole new set of challenges arise, because it may
be necessary for specialized teams of experts to work on
the biological assessments.

The maintenance of pipelines generally poses fewer problems along the majority of the corridor. Access to pump and
valve facilities are required. But the majority of the corridor can be revegetated and various land use alternatives
may be exercised, including agriculture. However, the
details will depend upon right-of-way agreements and other
legal constraints. From the biological perspective alone,
the potential to maintain a viable biotic community does
exist after the construction phase.

This discussion presupposes an underground pipeline.
Pipelines that include above-ground components (for example,
the trans-Alaska oil pipeline) have different maintenance
requirements. In fact, the trans-Alaska pipeline was sufficiently unique from an environmental viewpoint that a
special bibliography has been compiled by the U.S. Department
of the Interior (1974; see Selected References). References
on ecological, engineering and other aspects of the pipeline
are included.

SUMMARY OF POTENTIAL IMPACTS

Table 7 is a general summary of typical impacts commonly associated with utility corridors. Potential impacts
associated with the various utility corridors obviously
depend upon:

    type of utility
    length of corridor
    geographic location
    planning and design (including corridor selection)
      details
    construction practices
    maintenance practices

Thus there is considerable latitude to plan and manage the
construction of a utility corridor so as to mitigate potential, adverse impacts.

MITIGATION MEASURES

Many of the mitigation measures proposed for highway
projects (Chapter 14) also have relevance to utility corridors. Perhaps the most important mitigation measure centers on the selection of the corridor. Corridor selection
ideally ought to include considerable input from biologists
to assist in the evaluation of potentially sensitive areas.
However, both the engineer and biologist should realize that
many other inputs are required for an optimal selection process.
    Related to general corridor selection is the actual lo-
cation of the alignment. The alignment should consider
terrain features such as slopes, canyons, natural benches
and other topographic details. The general objectives are
to minimize cut, fill, and clearing for the right-of-way
and access roads. These measures will in turn minimize erosion, changes in hydrology, and loss of vegetation.
    Over the longer term, the most important mitigation
measure is to maintain the vegetation and thus the carrying
capacity for wildlife. Specific details, however, must
depend upon the objectives in the management and use of the
right-of-way.

TABLE 7. *Summary of Potential Impacts Commonly Associated with the Construction and Operation and Maintenance of Utility Corridors.*

Project phase	Potential impacts
Construction	Changes in surficial geology Changes in hydrology Erosion of soil Effects on water quality Loss of vegetation Loss and/or disruption of wildlife habitat Disruption of wildlife
Operation and maintenance	Use of pesticides may cause loss of vegetation and wildlife habitat Use of pesticides may cause effects on water quality and aquatic community Maintenance of cleared right-of-way may reduce carrying capacity Maintenance of cleared right-of-way may also increase species diversity Access roads may cause erosion channels Effects (largely unknown) on biota associated with very high voltages Collisions of birds with utility towers and power lines

The U.S. Department of the Interior in cooperation with the U.S. Department of Agriculture (1971) has published an excellent booklet on *Environmental Criteria for Electric Transmission Systems*. This booklet outlines many environmental criteria to use in selection, design, construction and maintenance activities for utility corridors. Among the criteria discussed are those related to minimizing the loss of wildlife habitat.

SOME ADDITIONAL COMMENTS

There are several other factors which are important from an environmental viewpoint, even though they are not primarily biological factors. A common complaint about utility corridors centers on adverse impacts on aesthetics. Tall transmission towers, power lines, poorly maintained corridors, and badly eroded access roads have all contributed a sense of intrusion on the natural landscape. Frequently opposition to transmission lines is based upon visual impact and not upon any specific environmental damage to abiotic or biotic components in the area.

Another factor is related to access roads and to the right-of-way itself. Frequently in densely forested areas the open right-of-way provides forage for deer and other game species. It also provides good access for hunters. Consequently the right-of-way corridor is a favorite area for hunters. This reality must be considered in any plans to enhance wildlife habitat.

There is considerable interest today in multiple uses of utility corridors. Potential uses include hiking trails, skimobile paths, equestrian paths, and other recreational uses. Also, as mentioned above, considerable potential for managed wildlife habitat exists for utility corridors. In some agricultural regions long transmission line corridors can take up significant acreage of land. Today there is considerable interest in the use of some of this right-of-way land for agricultural productivity. Conflicts in objectives and institutional barriers certainly exist. However, there will be increased pressures for multiple uses of utility corridors, especially uses related to conservation and recreation.

Environmental implications associated with the acquisition and management of rights-of-way will increase significantly as we experience increased land use constraints. The American Right of Way Association, Inc. (AR/WA) as a professional organization has recognized this increase, and has developed a training manual and training course on *Environmental Considerations* (New England Research, Inc., 1976). The approach of this training program is to stress the interrelationships among the abiotic, biotic, and human components of various activities associated with rights-of-way, such as utility corridors. This approach underscores the systems approach, and places the roles of the engineer and the biologist in the context of total team efforts.

Chapter 16

Water Resources Development

Perhaps no other natural resource has played as critical a role as water in the historical development of humans. The early developments of agriculture, transportation, commerce and industry were historically linked to water resources. This was true both in the Eastern and in the Western Hemispheres. Among the earliest accomplishments in civil engineering were those related to the development of water resources.

HISTORICAL OVERVIEW

Water resources were developed quickly in colonial America. The nineteenth century saw the extensive development of dams for water power and other purposes. Under the original *Rivers and Harbors Act of 1899 (Refuse Act)* the Chief of Engineers, as authorized by the Secretary of War, was directed to prohibit the obstruction of navigable channels. Since then the Act has been amended many times, and today it plays a major role in regulating engineering activities in navigable waterways.

Over the decades other legislation was enacted to develop and conserve water resources, to prevent erosion, to control damage from flood, and to provide water for irrigation (see Chapter 2). The U.S. Army Corps of Engineers, the Bureau of Reclamation of the U.S. Department of the Interior, and the Soil Conservation Service of the U.S. Department of Agriculture have played key roles in these efforts. Since the enactment of NEPA in 1970, the various projects and programs sponsored by these and other agencies are now subject to the assessment of potential environmental impacts.

The current legislative framework also includes various provisions of the *Federal Water Pollution Control Act Amendments of 1972* and various other laws relating to environmental protection and the development and conservation of natural resources.

Sponsorship of water resources development projects today may come from Federal agencies, regional commissions, state governments, local governments and private organizations. Various permits and frequently Federal funds are required for these projects. Consequently the relevance of environmental legislation is important to many of these projects. Among the environmental components that require assessment are those related to both aquatic ecosystems and terrestrial ecosystems.

TYPES OF PROJECTS

The types of projects and programs related to the use of water resources may encompass a variety of human objectives, including the following:

    waterways for navigation
    increasing agricultural production
    reclamation programs
    flood control
    erosion and sedimentation control
    water supply
    cooling
    hydroelectric power
    recreation
    enhancement of wildlife habitat

A critical assessment of this list reveals that conflicts in objectives are almost inevitable. Many of the environmental issues that arise in water resources projects are based on conflicts in social objectives.

The engineering alternatives available to achieve various objectives are extremely numerous. However, there are some general engineering approaches that are used extensively, including the following:

construction of dams and the creation of impoundments
channelization
out-of-basin diversions
dredging

Potential impacts associated with these approaches will be
discussed in the following sections. However, the construction of dams and the creation of impoundments results in so
much interest from diverse groups that most of the discussions in this chapter are centered on such projects.

IMPOUNDMENTS

The construction of a dam and the creation of an impoundment have the potential to create significant impacts
on both the abiotic and biotic components of the region.
Changes to the land mass, changes to the water mass, extensive construction activities and potential long-term effects from the operation of the impoundment are all important
factors to consider. Impacts on habitat, ecosystem changes,
changes in water quality, and various downstream changes can
all occur.
One of the major impacts associated with the creation
of an impoundment is the inundation and subsequent loss of
habitat for terrestrial wildlife. The aquatic habitat of
the previous stream channel which becomes part of the impoundment will also be changed. The old stream ecosystem
shifts to that of the lacustrine ecosystem (see Chapter 6).
The new system will be governed by the physical and chemical
characteristics of the new impoundment. For example, area,
volume, depth, morphometry, nutrients and watershed runoff
will all be important. However, regardless of abiotic
details, it is possible to predict that the community structure of the new aquatic ecosystem will change significantly
when compared with the previous stream ecosystem. Both
species diversity and density of primary producers and
various consumers (e.g., benthic organisms, fish populations)
will change. Because of the large increase in water mass
in the new impoundment relative to the original stream
channel, there is an increase in overall biomass of the
aquatic community. However, whether this is "good" or "bad"
depends upon the social objectives of the human community.

Loss of terrestrial habitat can be assessed by quantitating the area within the contour of the normal elevation of the impoundment. However, current regulatory guidelines require careful assessment of not only the quantity of habitat but also the quality of the habitat, including an estimate of carrying capacity and the presence of critical habitat. Also there will be considerable interest in impacts to the riparian habitat (see Chapter 6) along the old stream channel. Thus considerable effort is needed to document the existing terrestrial habitat that will be inundated.

Impoundments also cause changes in water quality. Water with heavy loads of suspended solids will actually be improved in quality because impoundments tend to retain water and thus allow materials to settle out. Relatively deep impoundments will undergo thermal stratification, and thus a lower layer of relatively cold water will exist. If the system is oligotrophic, this colder layer will be good habitat for coldwater fish (e.g., trout). If the system became highly productive and eutrophic, the lower layer will usually experience oxygen depletion, and thus it will be poor habitat for fish. Thus the water quality that will result will depend upon the physical, chemical and biological characteristics of the impoundment and the watershed.

The dam creating the impoundment also acts as a barrier to fish migrations in both directions. It is important to document whether the stream includes migrations of *anadromous fish,* because special attention to such species is required. The dam may also cause major changes to the hydrology, to the ground water aquifers and to the riparian vegetation of the downstream reaches of the stream. On the other hand, if the original stream experiences drastic changes in river stage with normal hydrologic cycles, the creation of the impoundment may actually provide flood control and low flow augmentation, if it is operated in that manner. In that way the downstream aquatic community may actually be partially stabilized against the disruptions of flood stages and low flow periods.

It becomes clear that the potential impacts to the regional ecology from the creation of an impoundment can be significant, and that the precise impacts will depend upon numerous details including, but not limited to:

surficial geology   aquatic biota
topography   design details
watershed details   construction activities
hydrology   operational details
terrestrial biota   human objectives

It is also clear that an adequate assessment of these potential impacts will require an interdisciplinary team
approach, as has been stressed repeatedly in this book.

CHANNELIZATION

Channelization is an engineering approach used to
achieve various objectives such as (1) drainage for land
reclamation, (2) flood control, (3) erosion control, (4)
improvements to navigation, and (5) relocation of a stream
for a project such as a highway.

The channelization may involve modifications to an
existing channel, it may involve the relocation of an existing channel to a new alignment, or it may involve the
creation of a channel where none existed before. Thus, the
potential impacts will vary considerably with the engineering
details.

Channels for drainage will cause changes in surface
water levels, changes in ground water levels, loss of wetlands, and loss of aquatic habitat. In fact, for many
decades the specific objectives of many reclamation projects
were to drain swamp lands (see Chapter 8). Filling in an
old channel and relocating the new channel in a rechannelization process will cause losses to both aquatic and terrestrial biota. However, depending upon details, the loss in
habitat may only be temporary.

Channel improvements and channel rectification of a
meandering channel frequently results in criticism and conflicts. Poor bank stabilization with subsequent erosion may
occur. Also channel rectification frequently results in loss
of tree and shrub cover along the previously meandering
channel bank. Cutting and dredging may result in spoil and
dredged material that has to be disposed nearby. There is
frequent criticism that the resulting channel, although well-
designed from a hydraulic perspective, has caused a loss in
aesthetic value.

Again the concept of human objectives is important. Reclamation of poorly-drained land can result in increased agricultural productivity. Agricultural land frequently provides terrestrial wildlife habitat. Also, the loss of swamp land results in the reduction of habitat for mosquitoes. Thus public health implications may also be associated with the drainage of wetlands through channelization.

OTHER ENGINEERING PROJECTS

There are numerous other engineering projects related to water resources, and a complete review is beyond the scope of this chapter. However, there are several types of engineering projects which frequently receive considerable interest from the public and from regulatory and natural resources agencies.

Among the water resources projects that receive considerable attention are those involving the *diversion of water* from one watershed to another. Out-of-basin and interbasin transfers of water involve so many physical, social, and institutional factors in the total human environment that close scrutiny is highly predictable. Outstanding examples on a national scale include major water supply projects for such metropolitan areas as Los Angeles, Denver, New York and Boston. Close scrutiny results in the need for careful assessment of any impacts that may result from such projects. Biologists add their efforts to such assessments. Efforts may include monitoring programs for projects already operational, and baseline studies and predictions of impacts for projects at the planning and impact assessment stages. Dams for impoundments, channels, aqueducts and water pipelines are all engineering components of such projects. Therefore all the potential impacts associated with the construction and operation of such engineering components may be involved. Another factor that is frequently involved is the potential introduction of undesirable species from one water resource to another. Also important is the careful evaluation of water quality changes that may result to the donor and/or receiver system.

Another engineering practice that has received much
interest in recent years is *dredging,* commonly in relation
to channel or harbor improvements for navigation. Many of
the waterways subject to dredging are often in industrialized and urbanized areas. Consequently, the mud and sediment often contains various potentially toxic substances
(e.g., heavy metals) that are sequestered in these bottom
deposits. Dredging will often release these substances in
the water column. Dredging also causes short term increases
in turbidity. A major environmental issue associated with
dredging is the disposal of dredged material. Various alternatives exist, including disposal on land or in water
(see Chapter 17). The concern for potential impacts related
to dredging is significant, and the Waterways Experiment
Station (WES) of the U.S. Army Corps of Engineers has a
whole program devoted to this important issue.

Hydroelectric power is another area where careful
assessment of potential impacts is required. Intake structures for the intake shafts leading to the power generation
units often result in the impingement of fish. Such *impingement* phenomena can lead to losses of fish against the screens
for the intake structures. Also, the mechanical *entrainment*
of fish eggs and larvae in the water column that passes
through the intake shafts and the turbines results in injury
and death. These biological effects receive considerable
attention and much effort is devoted to their careful analysis. ,More will be said about these effects in Chapter 18.

The use of *water for cooling purposes* is another area
where careful analysis of potential impacts is required.
Again, this assessment may involve both facilities that are
operational and those that are planned. The generation of
electric power results in most of the cooling demand.
Various cooling alternatives are used today, including once-
through cooling with rivers and cooling lakes, and off-
stream cooling with cooling towers. All alternatives involve the assessment of biological impacts. A substantial
portion of professional efforts in environmental biology
today are related to steam electric power plants and their
environmental effects. This general topic will also be
covered in greater detail in Chapter 18.

SUMMARY OF COMMON IMPACTS AND ENVIRONMENTAL ISSUES

The environmental issues that may be raised in relation to the development of water resources are extremely diverse. Also, the origins of these issues are diverse, and usually reflect the concerns of many individuals, organizations and agencies. The following list summarizes some common environmental impacts and issues raised in relation to these projects:

effects on water quality (general)
erosion/turbidity
depletion of oxygen
temperature effects
hydrologic changes
loss of terrestrial habitat
loss of aquatic habitat
loss of unique or critical habitat
threatened and endangered species
effects on wetlands
impingement of fish
entrainment of fish eggs and larvae
effects on fish migration
public health matters
toxic substances

Note that these various items involve both abiotic and biotic factors. The significant impacts that require careful attention will be highly project- and site-specific. However, the probability is very great that a project of any magnitude will require the careful assessment of most of the items on this list. Also, most of the mitigation measures (see Chapter 19) will center on the impacts and issues summarized on this list.

At the beginning of this chapter we stressed the importance of human objectives. As the demand for water resources increases, the potential conflicts among various objectives will also increase. The National Water Commission has completed an important study on *Water Policies for the Future* (1973). A repeated theme in this excellent book is the need for better development and management of water resources, along with sound environmental controls. The stage is clearly set for many years of future cooperative efforts among biologists, engineers, planners, and other professionals.

Chapter 17

Waterways and Marine Facilities

In this chapter we will examine such facilities as inland waterways, coastal waterways, and various marine facilities. The maintenance of navigation through dredging operations has important environmental implications. Recent years have brought increased pressures for the construction of deepwater ports. These deepwater projects in turn create development pressures for coastal areas. Public interest in all these activities has increased significantly in the past decade. Thus engineering activities related to dredging and to the construction and operation of marine facilities today receive careful assessments of a variety of potential impacts on both the abiotic and biotic environments.

IMPROVEMENT AND MAINTENANCE OF NAVIGATION

The United States has thousands of miles of waterways that carry commercial navigation. The Mississippi River Basin alone encompases some 9,000 miles of waterways from the upper reaches of the Mississippi and Ohio Rivers to the Gulf of Mexico. The Great Lakes and the coastal areas add extensive components to the waterway system of the Nation. Much engineering effort is invested in the improvement and maintenance of this system.

Dredging is a major activity used in the improvement and maintenance of navigation channels. It is estimated that the U.S. Army Corps of Engineers engages in the dredging of about 10 million cubic yards of materials on an annual basis for navigation purposes. Regulations administered by the Corps of Engineers and by the U.S. Environmental

Protection Agency include specific reference to dredged materials.   Much of the dredging is done in harbors and waterways in urbanized and industrialized areas.   Consequently, the physical acts of dredging and subsequent disposal may release potentially toxic materials from the sediments. These materials may include hydrocarbons, heavy metals, and other organic and inorganic compounds (see Chapter 11).   In addition, the dredging operation may release nutrients which can enhance primary productivity.   Finally, the act of dredging causes localized, short-term increases in turbidity. The biotic effects of increases in turbidity will depend upon the nature of the suspended particles.   In general, however, any effects on benthic organisms and fish are temporary.

The extent of dredging operations is carefully specified in an authorization for navigation improvements.   Consequently, channel length, channel depth and anchorage areas are specified.   The area to be dredged is generally small in relation to the total benthic area.   Within the area to be dredged there is generally a total loss of benthic organisms. The relative magnitude of this loss is related to the fraction of the benthic area that is dredged.   After dredging operations, benthic organisms will usually colonize the dredged area within a matter of months.   However, the nature of the new substrate may result in a different benthic community.

Disposal of dredged material is a subject of intense interest for many groups.   As mentioned in the previous chapter, the Corps of Engineers has an extensive research program devoted to the disposal of dredged materials.   There are three general approaches used for the routine disposal of dredged material:

1. *Disposal on Land*. This approach is based on the disposal of dredged material at a selected site on land. Usually a containment structure is included to prevent runoff of materials.
2. *Disposal in Shallow Water*. Disposal may also occur in relatively shallow water such as a bay or a river flat. The disposal area may be surrounded by a dike to retard exchange of water and the spread of dredged material.
3. *Disposal in Deeper Water*. Much of the dredged material is disposed in deeper water, including the ocean. However, ocean dumping is now restricted and may even be prohibited under certain conditions.

These various disposal operations may have impacts on biota
because of the release of particles and substances in the
water. Also disposal often results in the direct covering
over of benthic organisms. Dumping may also increase productivity, and often fishermen have found disposal grounds
at sea to be productive areas. However, the potential harm
from the accumulation of toxic substances in food webs is
now widely recognized.

Dredging operations are conducted in waterways that
include freshwater, brackish water, and salt water. A wide
spectrum of aquatic organisms and waterfowl may be associated with these waters. If the waters include commercial
fishing, special attention is required to assess potential
impacts on shellfish and finfish. Recreational use of water
is another important factor. Thus a wide range of potential
impacts, including impacts on aquatic wildlife, terrestrial
wildlife, water quality, commercial fishing, swimming,
boating, and aesthetics and the interrelationships among
these impacts, are frequently assessed by the project team.

HARBOR FACILITIES

Various harbor facilities may include breakwaters, docks,
piers, warehouses and ship repair facilities. Construction
and installation of these facilities will have localized
abiotic effects. These include increases in turbidity,
changes in various hydrodynamics (see Chapter 7), changes in
solar radiation, and changes in substrate. In addition,
navigation improvements such as dredging may be carried out
both as part of the initial project and periodically for
maintenance purposes.

Biotic effects frequently associated with such facilities include the direct loss of natural habitat. Of
special interest are highly productive areas such as
estuaries. Other concerns include the potential loss of
shellfish beds, potential impacts on fish spawning areas,
and potential impacts on shore birds and waterfowl.

Of interest is the excellent substrate frequently provided by piers in productive waters. Consequently, algae
and other organisms may develop extensive growths on such
substrates. These growths may provide food for consumers,
but they may also be considered an aesthetic nuisance.

Other impacts may include the discharge of various
potentially toxic chemicals and wastes in the water from
the various facilities. Harbor facilities may also have
secondary impacts on the aquatic environment because of the
increase in human activities generated by such facilities.
This is discussed further in a subsequent section.

DEEPWATER PORTS

The need for deepwater ports is a recent development,
and has resulted from the construction of very large tankers.
Ships in the size range of 200,000 to 400,000 tons are now
very common; such ships have navigational depth requirements
of 60 to 80 feet. Even greater depths are required for
ships over 400,000 tons. Thus the feasibility of conventional harbor facilities for such ships is sharply limited.
Deepwater ports at various distances from shore provide the
necessary facilities for such giant ships.

Deepwater port licenses are covered by the *Deepwater
Port Act of 1974* (see Chapter 2), and are administered by
the Department of Transportation through the U.S. Coast
Guard. An extensive environmental analysis is required as
part of the procedure for issuing a license for these ports.
The procedure requires a comprehensive assessment of the
potential environmental impacts associated with the construction, operation, and termination of deepwater port facilities.

The Coast Guard has prepared a *Guide to Preparation of
Environmental Analysis for Deepwater Ports* (1975). The
Guide specifies the types of baseline studies needed for
terrestrial and aquatic ecosystems. Of particular interest
is the definition of "important" species in this Guide; a
species is important:

1. If it is commercially or recreationally
    valuable.
2. If it is threatened or endangered.
3. If it affects the well-being of some
    important species within criteria 1 and
    2 above.
4. If it plays an important role in maintaining
    the structure and function of the ecological
    system.

5.   If it is an indicator species, such as clams or
     oysters, known and recognized for their easily
     recognizable reactions to pollutants or other
     adverse environmental impacts.

Application of ecosystem concepts (see Chapters 4 - 7)
tells us that many different species at various trophic
levels can be defined as important under this definition.
Although this is a Guide from one agency for one type of
project, the same definitions are frequently applied under
various guidelines for other types of projects.

   Another reason why this Guide is of interest is because
it is an excellent example of the application of systems
concepts. The language of the Guide is very clear in
specifying ecosystem concepts. The components and dynamics
of ecosystems, including various trophic levels, are described. Moreover, the marine-terrestrial interface environment is also stressed. Finally, the Guide also recommends
that the assessment of the whole system should include
humans as an important component.

   There are a wide spectrum of engineering activities
associated with deepwater ports, including dredging, the
construction of mooring facilities, the construction of onshore facilities, and the construction of pipelines.
Operations include the mooring of ships and the operation
of pipelines and other onshore facilities. Discharges from
ships may include ballast water, tank cleaning residue, and
sanitary wastes. Potential abiotic impacts associated with
these activities include:

   noise and vibration
   air quality changes
   water quality changes
   release of toxic materials and solid wastes
   changes in hydrodynamics
   oil spills

Among the potential biotic impacts associated with deepwater
ports are the following:

changes in the distributions of flora and fauna
interference with migratory movements
impacts on nesting, breeding and nursery grounds
effects on species-habitat relationships
environmental stress, especially from toxic
  materials and oil spills
loss of commercial fishery resources including
  shellfish and finfish
loss of terrestrial habitat
impacts on shore birds and waterfowl

An important consideration in the environmental assessment of deepwater ports centers on the terrestrial environment. Often forgotten are the extensive facilities required onshore. Therefore environmental impact assessments should include the coastal zone as well as the marine environment.

THE COASTAL ZONE

Deepwater ports, harbor facilities, and navigable waterways must be viewed as part of a vast network of other facilities. Deepwater ports may require extensive fuel storage facilities on the coast. Oil refineries may also be located nearby, frequently with a complex of petrochemical plants. Oil storage and handling facilities are complex, and the potential for system failures and oil spills must be recognized.

Coastal areas also require major transportation systems. Warehouses, grain elevators and oil and coal storage and handling facilities in relation to pipelines and to truck and rail facilities, are all part of this network. All these facilities and systems require extensive area. Consequently there is continual pressure to convert land use from natural areas to highly developed areas.

There is also an increasing demand for electric power. The generation of electric power at coastal sites is increasing significantly. Modern facilities, especially nuclear power plants, require large installations. Engineering components may include docking facilities, railroad facilities, fuel storage facilities, cooling water intake channels, cooling water discharge channels, smoke stacks and cooling towers. Thus transportation, navigation, and cooling water supply are important considerations with power plants.

The interface between the land mass and the water mass is often an area of great importance, and large power plants may have a wide range of impacts. These will be discussed in greater detail in the next chapter.

An important conclusion is that impact assessments on a project-specific basis may not provide the necessary overview on a larger, geographic scale. Individual land transportation, navigation, and marine projects focus on the specific requirements for approval and implementation. In recent years it has been recognized that comprehensive planning and regulation is also required. An important step in that direction is seen in the *Coastal Zone Management Act of 1972* (see Chapter 2). Systems engineering will play an increasingly important role in the analysis of alternative approaches to optimal land use in coastal areas. Thus navigation and marine facilities in the future should be considered in this broader framework.

Chapter 18

Energy Production

INTRODUCTION

The relationships between energy production and environmental considerations now receive considerable attention from public agencies, from private companies and from the engineering profession. The development and production of energy have become highly machinery-intensive during the past few decades. Moreover, the various energy-related activities are now conducted on very large scales. Consequently, their potential to cause environmental degradation has increased.

Strip mining operations, offshore drilling operations, and the construction and operation of large steam electric generating facilities all have the potential to cause major environmental effects on air, water and land resources. They also have enormous capacity to cause environmental effects on the biosphere.

The development and production of energy resources are highly regulated, and are subject to many of the major acts of environmental legislation (see Chapter 2). The Clean Air Act, the Clean Water Act, the Atomic Energy Act, the Resource Conservation and Recovery Act, the Surface Mining Control and Reclamation Act, and NEPA are all important examples of the legislation that has direct relevance to energy production. All components (air, water, land) of the abiotic environment receive careful assessment. However, it is the deep concern for potential impacts on the biosphere that has generated so much research activity in environmental biology. The health and environmental effects

related to energy production are a vast subject, and no
attempt is made to cover all aspects of the subject in
this chapter. The following topics have been selected in
an attempt to provide an overview of some of the current
environmental issues related to energy development and
production.

MINING OPERATIONS

The mining and processing of energy resources have
undergone considerable advances in technology in recent
years. One consequence is an increase in the scale of
various operations. The capacity to cause major effects
on land and water has increased in direct relationship
with the increase in scale of the operations. Development
of energy from coal, uranium, oil shale and oil sands may
be cited as examples of such operations.

Strip mining for coal, especially in the western states,
has become a major concern for many environmentalists.
Removal of the geologic overburden to gain access to coal
causes destruction of terrestrial habitat. Mining operations also cause contamination of water supplies, and
adverse effects on aquatic habitat can occur. Large
amounts of spoil accumulate, thus presenting the potential
for wind and water erosion. Much of the technology developed in recent years is centered on land reclamation
measures. Among the objectives are the refilling and
reclamation of the mining sites as soon as possible. Considerable progress has been made in the development of
revegetation techniques. Successful revegetation is important for the restoration of wildlife habitat. There is
opposition to reclamation because of the high costs.

The mining and processing of uranium ore also present
environmental concerns. Many of the activities related to
the mining, processing and milling of uranium are controlled
by the Atomic Energy Act. Public health and safety are
major concerns in the uranium industry. Radiation levels
from uranium ore are relatively low. Nevertheless, the
potential hazards from contamination with radionuclides
dictate the application of many protective measures. For
example, radioactive salts can leach from uranium ore and
from tailings and thus enter water supplies. Radionuclides

in water in turn may enter the food web.

More recently, the increases in energy costs have made other technologies more economically feasible. The mining and processing of oil shale and oil sands have become feasible alternatives. Both above-ground and *in situ* processing are possible. Major environmental problems include the potential contamination of water supplies and the need to dispose of large amounts of solid waste. Above-ground operations also have the potential to disrupt wildlife habitat. Reclamation measures become necessary to restore habitat.

OFFSHORE DRILLING

Drilling for gas and oil on the continental shelf underlying offshore waters (see Chapter 7) has increased in scope during the past two decades. Both the operation of exploratory drilling rigs and the construction of anchored platforms for oil production have increased in scope. These practices can cause impacts on marine ecosystems. However, it is the potential contamination of marine ecosystems with hydrocarbon chemicals that provides major environmental concerns from offshore drilling. Current areas of concern include the Gulf of Mexico, the coastal waters off the North Atlantic states, and coastal waters off California and Alaska.

Petroleum is a complex mixture of many hydrocarbons, including straight-chain aliphatic compounds and cyclic aromatic compounds (see Chapter 11). Contamination with petroleum hydrocarbons may result from major spills and blow-outs. Although blow-out prevention technology has advanced in recent years, blow-out episodes do occur. Blow-outs and spills of various types cause contamination of the water and littoral areas. These are the familiar oil contamination problems frequently publicized by the news media. Aquatic animals and shore birds become covered with oil and many succumb. Aesthetic problems add to the sense of drama, especially when oil contamination is widespread. Clean-up measures for spills have become important pollution control activities.

Oil spills during the 1970s have resulted in the accumulation of considerable data on the effects of petroleum
hydrocarbons on marine organisms. The American Petroleum
Institute, the National Oceanic and Atmospheric Administration, the U.S. Environmental Protection Agency and other
organizations have supported research programs on the
effects of oil spills. Studies are being done in the field
and in the laboratory. All biota, including plankton,
benthic organisms, nektonic species, birds and mammals are
being investigated. Much remains to be learned about the
various effects of hydrocarbons on biota. However, there
is evidence that these chemicals can accumulate in food
chains, and consequently even very low levels in the water
may produce harm. Thus the potential harm from chronic
exposure to low levels of petroleum hydrocarbons is now
receiving considerable attention.

POWER PLANT SITING AND CONSTRUCTION

The siting of power plants raises many environmental
issues, and much effort is needed for a comprehensive
assessment of various alternatives. Land, air and water
resources and their potential degradation are extremely
important and many environmental laws are directly relevant to these issues. The logistics associated with the
transport and handling of fuel are often of great magnitude.
For example, over its lifetime (approximately 35 years) a
large, coal-fired plant will require millions of tons of
fuel and will produce millions of tons of solid waste.

Land requirements for large steam electric power plants
may include thousands of acres of land, although the exact
requirements will vary considerably. Noise buffers,
cooling lakes, solid waste disposal sites, and fuel storage
facilities may all add to land requirements. Land use
changes, therefore, are inevitable. Relationships to
existing and additional transmission lines are also important considerations. These changes in land use have
direct relevance to the human environment. However, from
the biological perspective they will also have direct
relevance to wildlife habitat. Therefore, one of the major
efforts of biologists in power plant siting studies is to
conduct careful surveys of various habitats at the alternative sites.

Since new steam electric power plants will require
various permits, they are subject to the environmental
impact assessment procedure under NEPA. Federal agencies
will vary, and may include the Rural Electrification Administration (REA) of the U.S. Department of Agriculture, the
Nuclear Regulatory Commission (NRC), and the U.S. Army
Corps of Engineers. Each agency has different regulations,
guidelines and policies that stress various factors. Siting
studies eventually lead to the comprehensive impact assessments of viable alternatives as required under NEPA. Among
the assessments will be those related to the abiotic elements and the biotic elements of the proposed sites.

The construction of steam electric power plants and
associated facilities requires several years. Therefore
many of the impacts usually associated with the clearing
of extensive land areas and the subsequent construction of
various facilities will occur. The construction of cooling
lakes, railroads, transmission lines, intake and discharge
canals, and the main power plant structures is a major
undertaking, and many of the abiotic and biotic impacts
reviewed in other sections of this book may take place.
Much of the concern over the environmental effects of
power plants, however, is related to the operation of the
plants.

STEAM ELECTRIC POWER GENERATION

A major environmental concern associated with the
generation of steam electric power centers on the discharge
of heated water from the condenser cooling system. Cost-
effectiveness has recently dictated the need to build
relatively large power plants (e.g., over 1000 MW generating capacity). Such plants use large amounts of condenser
cooling water. The design of the condenser cooling system
has important implications to the aquatic organisms in the
water bodies from which the cooling water is drawn and
into which it is discharged.

The water passing through the condenser cooling system
may be heated by 10-20 degrees F (or more) above ambient
temperature. The greater the heat differential (ΔT), the
greater the potential harm from the thermal effluent. Once-
through flow of the condenser cooling stream may be based

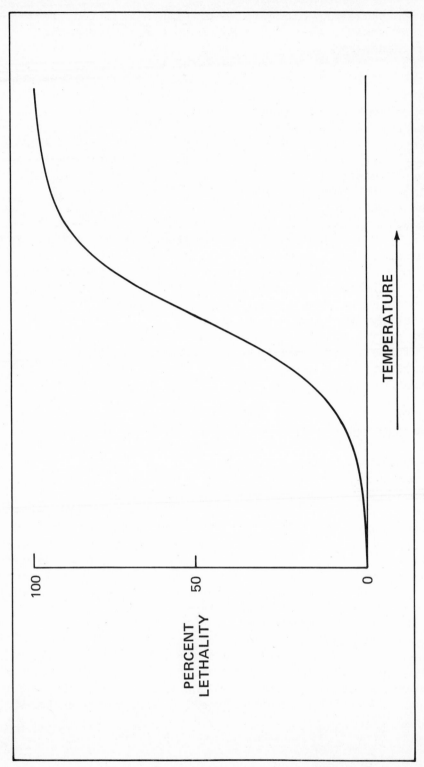

FIGURE 16. *Stress-Effect Curve Showing the Relationship Between Temperature Increase and the Probability of Lethality for Aquatic Organisms.*

on the use of a river or a cooling lake. The heated effluent is generally discharged as a thermal plume which
eventually cools to the ambient temperature. However, as
the heated water goes through various temperature changes,
these temperatures can have different effects on various
biota (e.g., plankton, fishes). Figure 16 illustrates
the stress-response relationship (see Chapter 10) between
temperature and mortality for aquatic organisms. Note
that death by heat stress is a probability concept, and
what happens to fish populations is governed by the fish
species, the temperature and other factors. The investi-
gation of the effects of heated effluents has been an
active area for research by environmental biologists. These
investigations are governed by Section 316(a) of the
Federal Water Pollution Control Act Amendments of 1972;
they are referred to as 316(a) Demonstration studies.

   Closely related are studies required under Section 316(b)
of the Federal Water Pollution Control Act Amendments of
1972. The emphasis now shifts to potential mechanical
damage to aquatic biota. *Impingement* losses come from the
inability of fish to avoid being carried against the intake
screens by the inflow of cooling water into the intake
structure. As a general rule, most of the fish lost
through impingement are small (5-10 cm). Mechanically held
against the screens by the flow of water, small fish can
suffer mechanical damage. Larger fish generally are
powerful enough swimmers to escape. Many factors will
influence impingement losses including (1) fish species,
(2) schooling behavior, (3) temperature, (4) oxygen levels,
(5) swimming speed, and (6) physiological condition of the
fish.

   Smaller organisms readily pass through the screens with
the water mass, and thus become entrained in the cooling
water. Small, microscopic and generally spheroidal biota
such as algae suffer relatively less damage from *entrainment*
than the larger biota. Probably on the order of 75 to 90
percent of the phytoplankton remain viable after entrainment. On the other hand, larger biota such as fish larvae
(several mm long) may suffer close to 100 percent mortality
during the entrainment process. Cause of death is usually
a combination of both thermal and mechanical effects.

The thermodynamics of cooling are such that the heat increase in the thermal discharge can only be reduced through an increase in flow rate. Thus there is a potential conflict between measures to reduce the temperature of the thermal discharge and measures to reduce the flow of cooling water. In that sense 316(a) and 316(b) considerations will also conflict. Thus an optimal balance to minimize adverse impacts requires careful assessment of many factors in the entire aquatic ecosystem. Engineering data, thermal data, hydraulic data and biological data must be integrated and evaluated in a systematic manner on a site specific basis. The U.S. Environmental Protection Agency has issued several guidance manuals for 316(a) and 316(b) studies (see Selected References).

OTHER ENVIRONMENTAL CONSIDERATIONS

Nuclear power plants present the potential to introduce radionuclides into the environment. Generally, the radiation levels released from nuclear plants in good operating condition are minimal. Nevertheless there is a constant potential for release of radiation, and constant monitoring and surveillance are required.

More critical is the widespread concern with the safe disposal of nuclear wastes. Although several technical approaches to the disposal of nuclear wastes appear feasible, no alternative has met with universal acceptance by government, industry and environmental groups.

Fossil-fueled power plants release sulfur dioxide into the atmosphere, and these emissions can cause damage to plants and animals. They can also cause respiratory disorders in humans. The widespread phenomenon of *acid rain* is due to sulfur and nitrogen oxides which combine with moisture to form acids in the atmosphere. The sources of these contaminants include power plants.

Coal-fired plants can release other contaminants into the environment. A large coal-fired plant can release more radionuclides than a nuclear power plant. Coal contains significant amounts of radioactive materials. Coal-fired plants also release traces of other contaminants such as heavy metals into the atmosphere, even when equipped with scrubbers. These trace contaminants are cycled throughout the environment and eventually enter the biota.

TABLE 8. *Summary of Major Environmental Considerations*
Related to the Development and Production
of Energy.

Energy-related activities	Major environmental considerations
Mining operations	Disruption of terrestrial habitat Contamination of surface water and groundwater Disposal of solid waste and land reclamation
Offshore drilling	Disruption of aquatic ecosystems during exploration and construction Contamination of water column and littoral zone with petroleum hydrocarbons
Power plant siting	Requirements for adequate land and water resources Effects on wildlife habitat Effects on air quality and water quality
Operation of nuclear power plants	Thermal effects on aquatic biota Entrainment and impingement of aquatic biota Release of radionuclides in the environment Disposal of nuclear wastes
Operation of fossil-fueled power plants	Thermal effects on aquatic biota Entrainment and impingement of aquatic biota Release of sulfur dioxide Release of radionuclides and trace metals in the environment

SUMMARY

Energy-related engineering activities include extraction, processing, and other operations. These activities now have enormous capacities to influence the atmosphere, the hydrosphere and the lithosphere. They also cause direct and indirect effects on the biosphere. Much of the current research in environmental biology is centered on the effects of energy-related projects. A summary of major environmental effects related to energy development and production is shown in Table 8.

Chapter 19

Mitigation Measures

Mitigation measures applied to adverse environmental impacts from engineering projects are now receiving much attention. Legislative support for the application of mitigation measures to environmental impacts is rather widespread. Perhaps the greatest impetus, however, has come from the implementation of NEPA, where various agency guidelines and regulations have included provisions for minimizing adverse environmental impacts.

The precise language will often differ, and the various regulations make frequent reference to the words "ameliorate", "minimize", "mitigate" and "enhance". In this chapter the term "mitigation" is used to denote various actions which might be taken to minimize adverse impacts on the regional biosphere.

GENERAL APPROACHES TO MITIGATION MEASURES

We can go back to our ecosystem concepts (see Chapter 4) for guidance on general approaches to mitigation measures. One general rule is that the project should be implemented with minimal disruption of the abiotic environment. Another general rule is to avoid direct losses of or impacts on biota. Loss of or changes in wildlife habitat should be avoided or kept to a minimum. And finally, major construction activities should be conducted during seasons in which there will be the least disruptions to the regional ecology.

Actual details and methods for the implementation of mitigation measures are generally centered on the following:

```
project planning considerations
project location considerations
project design considerations
management practices for the construction phase
management practices for the operation phase
management practices for maintenance activities
```

Careful attention to these various project activities can have highly significant consequences to the regional ecology.

THE TERRESTRIAL ENVIRONMENT

One of the most important mitigation measures may be in the *location* of the project site. The selection of highway corridors, transmission lines corridors, and sites for power plants, as examples, can have great significance to the terrestrial biota. Biologists working in close cooperation with engineers can provide valuable input in the location phase of engineering projects.

The *planning and design* phases can also be important in the mitigation of potential impacts on terrestrial ecosystems. The extent or magnitude of a project should receive careful consideration. Optimal use of terrain features is important. Especially important are design features which minimize cut and fill requirements. Another important measure is to minimize the clearing of vegetation.

Construction activities have been recognized for years as a major disruption of the terrestrial environment. Consequently, much engineering talent has been devoted over the years to the careful *management of construction* activities. Erosion control measures, grading, revegetation, and minimizing the extent of exposed areas are all important. Numerous guidelines and technical manuals have been published on these subjects (e.g., Erickson *et al.*, 1978; see Selected References). More recently there has been increasing concern about the potential disruption of wildlife populations during nesting and breeding periods. Consequently, construction activities are often curtailed during such sensitive periods.

Much of the mitigation effort today is devoted to minimizing the loss of wildlife habitat. In addition to measures taken during the location and design phases, it is often possible to create and preserve habitat diversity after construction. There is much interest in wildlife habitat enhancement measures, and more can be done in this area. Particularly important is the need to avoid critical habitat.

As pointed out in Chapter 11, there is an increasing interest in the potential effects of various chemicals in the environment. Pollution control features are provided in our technology today, and many problems are greatly minimized. However, there are still many activities that require careful attention. Today there is an increasing interest in minimizing the use of salts on the highways. Maintenance practices of rights-of-way by both private organizations and public agencies are moving in the direction of minimal use of herbicides. Design and maintenance practices to avoid spillage of chemicals in the environment are important. Careful selection of solid waste disposal sites are also important mitigation measures.

THE AQUATIC ENVIRONMENT

The planning, location and design of projects that interact with the aquatic environment also provide opportunity for the mitigation of adverse impacts. Projects should generally avoid lakes, streams, wetlands, and estuaries. Stream crossings by highways should incorporate good culvert design, not only for hydraulic requirements, but also for aquatic habitat. Rechannelization should be avoided; if it cannot be avoided it is possible to design the new channel with habitat diversity and quality to maintain a viable aquatic community. Obstruction of fish movements and migrations should be avoided in any project that encroaches upon a stream channel. Spawning grounds should also be carefully avoided.

Aquatic ecosystems can be influenced by changes in water quality, and many measures can be taken to mitigate potential adverse impacts. Runoff from erosion of land can increase turbidity, nutrients, and oxygen demand in the water. These events in turn may increase the stress on aquatic organisms. Toxic materials such as various biocides, herbicides and hydrocarbons, whether used on land or

water, have the potential to cause damage to food webs.
Therefore, erosion control measures, minimal use of pesticides, and the prevention and control of accidental spills
are important mitigation measures. Chemicals move easily
from air to land to water as they are constantly cycled in
the environment. Therefore, any divisions between terrestrial and aquatic ecosystems are arbitrary, since all
systems are dynamically linked (see Chapter 6).

Various publications, including practical guidelines,
are being published on various approaches to minimize
adverse impacts. Many of these approaches emphasize location selection and construction activities. Among the
organizations that have published such guidelines are the
U.S. Environmental Protection Agency, the U.S. Department
of Agriculture, the U.S. Department of the Interior and the
U.S. Department of Transportation. (See Selected References
for examples of such publications.)

SOME SPECIES ISSUES

There are certain environmental issues that are important enough to warrent emphasis in this discussion of
mitigation measures. In Chapter 9 the importance of fish
and wildlife resources were discussed at some length. Relatively large species such as fish and mammals receive considerable attention for the various reasons discussed
previously. Some potential adverse impacts and possible
mitigation measures associated with these species are
discussed below.

Major highways usually result in numerous cases of
roadkill of large mammals. Perhaps the most dramatic examples are cases of deer killed by collisions with vehicles.
In the state of Pennsylvania alone over 20,000 deer have
been killed annually on highways. Among the measures that
have been used nationally to reduce the loss of large
game are fences, overpasses, underpasses, reducing speed,
the use of signs to alert drivers, and the avoidance of
preferred habitat and migration routes. This is an area of
active research, and much more needs to be done to reduce
this loss of wildlife. The American Association of State
Highway and Transportation Officials (AASHTO) is active
in this area, and much of the effort in highway design is
related to the problem of collisions with large animals.

Fishery resources also receive close attention, and frequently become major issues. Among the more important issues are those related to the thermal effects and to the entrainment and impingement of fish from the operation of steam electric power plants (see Chapter 18). Reduction in thermal discharges into rivers by the use of cooling towers, cooling ponds and cooling lakes are various alternatives used today. Another area that is receiving interest is the use of low grade heat from power plants to increase biological productivity in aquatic systems. Various designs of intake structures have been adopted to minimize the losses of fish and aquatic wildlife from entrainment and impingement. Among the measures that have been adopted are reduction in the flow velocity of water through intake structures, the design of horizontal screens in place of vertical screens, and various devices to return impinged fish back into the water away from the cooling stream. Much of the current research effort in environmental biology is devoted to the assessments of thermal, entrainment and impingement effects, and potential measures to mitigate these effects. Among the organizations active in these efforts are the U.S. Environmental Protection Agency (EPA), the Energy Research and Development Administration (ERDA), and the Electric Power Research Institute (EPRI).

The mitigation of potential impacts on wetlands is another area of special interest. The U.S. Army Corps of Engineers, the U.S. Department of Transportation, the U.S. Environmental Protection Agency and National Cooperative Highway Research Program (NCHRP) have all conducted research programs related to engineering projects in relation to wetlands. Because of the legislative background, public interest, and biological uniqueness of wetlands (see Chapter 8) much interest is generated by projects that encroach on wetlands. Measures to mitigate losses or to create equivalent wetland habitat are commonly required. For many engineering projects the U.S. Fish and Wildlife Service is also required to evaluate the potential loss of wetland habitat, and the mitigation of any losses receives considerable attention.

Perhaps no issue in environmental biology represents as complex a combination of legal, aesthetic, philosophic and technical factors as the issue of threatened and endangered species (see Chapters 2 and 5). Original interest was centered on fish, birds and mammals. Now there are plants and invertebrate animals on the lists of threatened and endangered species. Large projects involving large areas

must be assessed carefully for this issue. The seriousness
of the legal basis for species protection became apparent
in 1976 when a court order stopped work on the Tellico Dam
project in Tennessee. This $100 million-dollar project was
80 percent completed, but the survival of a rare fish
species (the snail darter) was threatened by the project.
The proposed Dickey-Lincoln School hydroelectric project
in northern Maine may experience legal constraints because
of the discovery of a rare species of plant (lousewort) in
the project area. The critical habitat requirements for
rare organisms are not easily duplicated. Consequently
proposed mitigation measures based on the creation of
equivalent habitat may be opposed on technical grounds.
Each case involving critical habitat is unique, and many
factors must be considered to resolve the issue.

SOME ADDITIONAL COMMENTS

Many of the earlier proposals for mitigation measures
under NEPA were *ad hoc* responses to obvious problems.
Projects were often well along in the planning and site
selection phases. Today there is greater opportunity to
look at the project planning phases as an important mitigation step. Systems planning on a wide geographic scale
is becoming more and more important. Biologists and
engineers can both play major roles in this planning function.

Another important aspect of the mitigation of adverse
impacts on living systems is based on a better understanding
of the importance of the total biosphere to humans. This importance is based upon our interests in conservation, agriculture, fisheries, recreation and environmental health.
There is now increasing concern for the quality of the
environment. This concern rests on legislative realities,
on public interest in environmental issues, and on scientific data.

There is good reason to believe that engineers will be
under constant pressure to minimize or even eliminate the
risks of damage to biosystems through various measures.
Many groups have forgotten that good engineering practices
have always attempted to minimize impacts on the abiotic and
biotic environments. These practices have antedated the era
of NEPA and other legislation by many decades. However,
abuses of the environment have also occurred, often

dramatically apparent to the public. The environmental legislation has recognized these earlier abuses, and much effort is now devoted to mitigation measures.

An important contribution to the literature on mitigation measures for biological resources was published recently (Swanson, 1979). This volume summarizes a symposium on the state-of-the-art on mitigation measures, especially as they apply to wildlife habitat. This and similar publications can provide the engineering community with valuable guides on approaches to mitigating losses of biological resources. Mitigation measures should be based on sound scientific and technical information, and biologists and engineers working together have enormous opportunities for important innovations in the future.

Chapter 20

Some General Guidelines

The ultimate success of environmental studies will be determined in great measure by how well the individual disciplines are applied to the specific objectives of the project. Frequently individual disciplines will be represented with sophisticated studies, only to fail because they were not fully responsive to the objectives. In other instances individual studies are not integrated into the general conclusions, and thus their direct value to decision-making is limited. Some of these problems were reviewed in Chapters 12 and 13. The application of biology to environmental studies has also had its failings over the past few years. Excessive costs, poor communications with agencies and with the public, project delays, and resort to adjudicatory proceedings have been common experiences. Much has been learned from these past failings.

This book is addressed to a wide group of professional scientists and engineers, and therefore the assumption is made that many readers will not be directly involved in the performance of biological studies. A further assumption is made that this book will be useful to environmental engineers, environmental scientists, project engineers, and project managers. Thus most of the comments in this chapter are addressed to professional scientists and engineers who will routinely participate in, manage, or review complex environmental studies. It is important to note that this is not a book on how to prepare an EIS (see Chapter 2). Nor are these guidelines restricted to environmental studies conducted for projects subject to NEPA. Rather, these comments focus on various steps which are critical to the successful performance of biological components in environmental studies.

LEGISLATIVE AND REGULATORY CONTEXT

The biological team should be thoroughly familiar with the legislative background and regulatory context of the project. It might be argued that the individual scientists do not have to know the regulatory details. However, experience has shown that thorough familiarity with the legislative background results in highly relevant biological studies. The early years were full of examples where biologists conducted irrelevant studies for environmental projects. Although much progress has been made, there is still need to be alert to this issue. The principal investigator or project director for the biological studies, especially, should be thoroughly familiar with the appropriate regulations.

DESCRIPTION OF ENGINEERING DETAILS

The biologists should also be familiar with the engineering details and specifications of the project. This applies equally to cases of existing facilities and to proposed projects. Examples of engineering details include the following: lane-miles of highway; proposed grade crossings; pipeline specifications; proposed time frame for the construction phase; acre-feet of water volume; depth of water; maximum pumping capacity; and length of channel. Each project has its own details and specifications; most projects also have several alternatives. Maps and drawings are extremely useful to biologists. As the study proceeds, it is common practice for the biologists to request additional engineering data. The alert project manager should investigate cases where the biologists are not requesting information on engineering details; he or she may discover that their efforts are not being coordinated with the engineering aspects of the project.

SCOPE OF WORK

It is common practice for engineering projects to be based on a detailed scope of work. It has become common practice for biological studies, especially extensive studies, to also include a scope of work. At least three elements should be specified in the scope of work.

1. *Objectives*. The precise objectives of the
   biological studies should be outlined. It
   is good practice to make appropriate
   reference to the specific legislation and
   regulations which provide the background
   for the project.
2. *Technical Details*. The technical details of how
   the work will be conducted should also be
   specified. Particularly important are
   details on field methods, laboratory
   methods, statistics and the evaluation of
   data.
3. *Administrative Details*. The administrative
   details should include an estimate of the
   level of effort, the project timetable, and
   the types of reports.

A scope of work may be routine for contract services
for biological studies; however, it is not always prepared
for work to be done by the staff. Thus biologists working
for public agencies or utilities frequently do not have a
formal scope of work for work projects. It is good management practice to have a scope of work, no matter how brief,
if the work is considered important.

SPECIFICATION OF OBJECTIVES

The specification of precise objectives is very important for the efficient coordination of the overall
project. Objectives for biological studies generally include baseline studies, compilations of inventories of
biota, monitoring programs for existing facilities, the
assessment of potential impacts of proposed facilities,
specialized research projects, or any combination of these
possibilities. It is in the specification of objectives
that the legislative and regulatory context becomes important. Implicit in the objectives is the conceptual framework for the biological studies. The objectives provide
the rationale for the technical details of the scope of
work.

GUIDES TO TECHNICAL DETAILS

Technical details for the biological studies will rest on the following:

1. *Ecosystem Concepts.* Ecosystem concepts provide a strong conceptual base for the planning, the execution, and the interpretation of the results of biological studies. We are fortunate to have over 100 years of accumulated knowledge in ecology, and this theory base is valuable for the biological components of environmental studies.

2. *Agency Guidelines.* The legislative and regulatory context of the project also provides a guide for the technical details. Regulations, guidelines, and technical manuals relevant to the project objectives will provide valuable information. It is surprising how often these various documents from the relevant agencies are overlooked.

3. *Literature on Biological Methods.* The literature on the methods to conduct field and laboratory work related to biological studies is vast. However, certain publications have become standard references, and these will provide a guide to the biological studies. Many of these publications are included in the list of Selected References.

The biological studies specified in the scope of work should include reference to (1) the *parameters* that will be studied, (2) the *locations* where the parameters will be studied, (3) the *frequency* with which they will be studied, and (4) the *methods* that will be used. In essence these specifications answer the questions – *what, when, where* and *how?* Ecosystems are large, complex and dynamic, and every parameter, component, location, and day cannot be studied or monitored. By the application of ecosystem concepts, agency guidelines, and accepted methods, the technical details can be specified in the appropriate manner. It is important to note that biological studies may be too detailed and thus very expensive in relation to the project objectives. They may also be inadequate to describe the

environment in sufficient detail for a sound decision. Much
judgement is required for the optimal balance in the biological studies in relation to the total project objectives.
See Appendix 3 for additional comments on biological
studies.

Another important consideration is the acquisition of
environmental data. Frequently much information on the
regional environment is available from various sources such
as government agencies and universities. Such data as hydrology, meteorology, water quality and lists of plants and
animals are frequently available. A decision to use these
data can save considerable time and expense in environmental
studies. However, since much of this information has been
generated for various objectives, gaps may exist in the data.
Therefore, the decision on whether to generate additional
data must be reached carefully.

ADMINISTRATIVE MATTERS AND PROJECT MANAGEMENT

As mentioned above, it is good practice to include
administrative details in the scope of work for biological
studies.

1. *Level of Effort*. The worker-hours or worker-days
    of effort should be estimated based on the
    technical details. This is true for projects
    conducted under contract services and for
    projects conducted by the in-house staff.
2. *Timetable*. A specific timetable for the
    biological studies is important. Field work,
    laboratory work, reports and other work de-
    tails should be organized within a project
    timetable.
3. *Reports and Public Meetings*. Written and
    oral communications should also be specified.
    Progress reports are important in relatively
    long (multi-season) studies. Presentations
    at public meetings and hearings can be ex-
    tremely important, and these activities must
    be planned carefully.
4. *Liaison and Coordination*. Finally, the
    management of the project should include care-
    ful attention to liaison with outside groups.
    Generally it is desirable for biologists to
    contact technical personnel from various

agencies such as fish and wildlife
agencies. Also, the coordination of the
environmental team with frequent meetings
is important.

Environmental studies rarely fail to achieve their
objectives because of lack of technical expertise. Technical personnel are alert to the potential needs for specialized expertise, and consequently, the expertise is acquired.
However, the expertise must be applied effectively. The
most common cause of failure lies in poor project management,
liaison and coordination. This is especially true with complex technical projects with complex procedural and regulatory requirements.

IMPACT ASSESSMENT

Much of the work in environmental biology today is
related to the assessment of the impacts of human-caused
perturbations in the environment. These impacts may be
associated with existing facilities, or they may be potential impacts from future activities. The most complex assessments are generally those related to projects subject
to NEPA. In those cases, irrespective of project details,
the biological team can expect certain predictable elements
in their studies.

1. *Description of Existing Environment.* Whether
    through the acquisition of available data,
    through field work, or through laboratory
    efforts, the objectives are to describe the
    existing environment and any dynamic trends.
    Abiotic components, biotic components and
    habitats are studied, and a description of
    the dynamics of the regional ecology is developed.
2. *Proposed Actions and Alternatives.* Details and
    specifications of the proposed action and its
    alternatives must be considered by the
    biologists and other team members. Especially
    important are quantitative data which describe
    the construction and operation of engineering
    facilities.

3. *Prediction of Probable Impacts*. Prediction
    of the probable impacts of the proposed
    actions and the alternatives is the key
    step in impact assessment. Many methods
    have been developed and include system
    diagrams, matrices, and graphic overlays.
    Among the impact parameters that should
    be assessed are the probabilities, time
    frames, and magnitudes of the impacts.
4. *Relevance of the Impacts*. The impact assess-
    ment process is an aid to decision-making, and
    consequently the impacts should be assessed
    for their relevance to the local human en-
    vironment. Biological resources may
    have relevance to several local issues, and
    the project team must assess these issues.
    Recent guidelines, especially, tend to stress
    the relevance of impact assessments to local
    issues.
5. *Mitigation Measures*. Potential adverse
    impacts of significance to the local environ-
    ment may be singled out for further assess-
    ment. These assessments will frequently
    include measures to mitigate adverse impacts.
    There is great potential for innovation in
    such mitigation measures related to engineering
    projects.

COMMUNICATIONS AND DECISION-MAKING

Often forgotten during the detailed activities of a
project are the overall reasons for the biological studies
included in an environmental assessment project. Ultimately one or more *decisions* must be made. Whether to issue
a license, to construct a dam, to select a specific corridor,
to provide additional pollution control hardware, or to
mitigate potential losses of wildlife - these are some of
the decisions that are common. Consequently the results of
the biological studies must be *appropriate* to the decision-
making process. The description of the regional ecology
must be *accurate*. The assessment of the project impacts
must be *objective* and must reflect the state-of-the-art.
Also, the assessments must be highly *relevant* to the local
citizens and to the regulations. Finally, the information
must be *communicated* in such a way that all people involved

in the decision-making process can understand the implications of the various alternatives. This process will involve public agencies, which typically require technical details; it may also involve private citizens, who frequently object to technical details. Communications achieve maximum results only when they disclose fully and clearly all the consequences to alternative actions. The communication of technical information in the appropriate format is one of the major challenges facing biologists and all professionals involved in environmental assessments.

SUMMARY OF GENERAL GUIDELINES

In summary, the following list is offered as general guidelines to the successful application of environmental biology to engineering projects:

1. Develop a scope of work for the biological studies.
2. Clearly specify the objectives of the studies.
3. Identify the background legislation and regulations.
4. Make reference to agency guidelines, technical manuals and other pertinent documents.
5. Relate the biological studies directly to the engineering data and specifications.
6. Include studies of the alternatives, if applicable to the project.
7. Specify which biotic and abiotic parameters and components will be studied.
8. Specify the locations to be studied.
9. Specify when and with what frequency the field studies (surveys) will be conducted.
10. Specify what field and laboratory methods will be applied to the studies.
11. Indicate what will be included in the environmental assessment efforts.
12. Indicate the proposed project timetable.
13. Specify the proposed budget and/or level of effort to be expended for the studies.
14. List the interim reports and final report that will be produced in the studies.
15. Indicate what appearances and presentations will be made at public meetings and other citizen participation activities.

Over a decade of experience has indicated that careful application of these guidelines will greatly enhance the interdisciplinary efforts between environmental biology and engineering.

Appendix 1

Biological Nomenclature

The two major subdivisions of all living organisms on earth consist of the plant kingdom and the animal kingdom. Biologists have devised various classification systems for both the plants and the animals.

THE PLANT KINGDOM

Numerous classification systems for plants exist. One generally accepted system is as follows:

Subkingdom *Thallophyta* (simple plants without roots, stems and leaves; do not form embryos)
 Phylum *Cyanophyta* (blue-green algae)
 Phylum *Euglenophyta* (euglenoids)
 Phylum *Chlorophyta* (green algae)
 Phylum *Chrysophyta* (yellow-green algae, golden
   brown algae, diatoms)
 Phylum *Pyrrophyta* (cryptomonads, dinoflagellates)
 Phylum *Phaeophyta* (brown algae)
 Phylum *Rhodophyta* (red algae)
 Phylum *Schizomycophyta* (bacteria)
 Phylum *Myxomycophyta* (slime molds)
 Phylum *Eumycophyta* (true fungi)
Subkingdom *Embryophyta* (plants forming embryos)
 Phylum *Bryophyta* (mosses, liverworts, hornworts)
 Phylum *Tracheophyta* (plants with vascular tissues)
  Subphylum *Psilopsida* (psilopsids)
  Subphylum *Lycopsida* (club-mosses)
  Subphylum *Sphenopsida* (horsetails and
    relatives)
  Subphylum *Pteropsida* (ferns and seed plants)

Most of the common plants with which we are familiar belong to the phylum *Tracheophyta*. This group includes flowers, herbs, shrubs and trees. Other plants important in environmental issues include the various groups of algae (e.g., *Cyanophyta, Chrysophyta*).

THE ANIMAL KINGDOM

Numerous classification systems also exist for the animal kingdom, and many experts have proposed various systems. For our purposes, the following classification of major phyla is appropriate:

> Subkingdom *Protozoa* (unicellular animals)
>   Phylum *Protozoa* (amoeba, paramecia, ciliates)
> Subkingdom *Metazoa* (multicellular animals)
>   Phylum *Porifera* (sponges)
>   Phylum *Coelenterata* (hydra, jellyfish, corals)
>   Phylum *Platyhelminthes* (flatworms)
>   Phylum *Nematoda* (roundworms)
>   Phylum *Rotifera* (rotifers)
>   Phylum *Annelida* (segmented worms)
>   Phylum *Brachiopoda* (lamp shells)
>   Phylum *Mollusca* (molluscs, clams, snails)
>   Phylum *Arthropoda* (insects, spiders, crustaceans)
>   Phylum *Echinodermata* (starfish, sea urchins)
>   Phylum *Chordata* (tunicates, acornworm, vertebrates)
>   Subphylum *Vertebrata* (vertebrates)

Much of the interest in environmental studies centers on the vertebrate animals. Consequently, it is useful to list the subclassifications of this group; the vertebrates may be divided into five *classes* as follows:

> Class *Pisces* (fishes)
> Class *Amphibia* (frogs, salamanders)
> Class *Reptilia* (turtles, snakes)
> Class *Aves* (birds)
> Class *Mammalia* (mammals)

BIOLOGICAL NOMENCLATURE

The science of biology has developed an elaborate terminology to classify organisms into appropriate categories. Some of these terms (kingdom, phylum, class) have been listed above. It appears useful at this point to carry the classification of one species (the frog) through the entire hierarchy of nomenclature as an example:

Kingdom *Animalia*
    Subkingdom *Metazoa*
        Phylum *Chordata*
            Subphylum *Vertebrata*
                Class *Amphibia*
                    Order *Anura*
                        Family *Ranidae*
                            Genus *Rana*
                                Species *pipiens*

Thus in scientific nomenclature the common frog is called *Rana pipiens*.

The scientific names for all organisms will have a *genus* (generic component) and a *species* (specific component). In the biological literature the genus (e.g., *Rana*) is capitalized, and the species (e.g., *pipiens)* is in lower case. The scientific names are underlined in the typewritten format and generally italicized in the printed format.

The science of classification (taxonomy) is based on a system of Latin names. Each category of classification (i.e., class, order, family, genus, species) is called a *taxon*. Thus the lowest taxon is the species.

TAXONOMIC DETAIL IN ENVIRONMENTAL STUDIES

Environmental studies frequently require that biological organisms be identified to the lowest practical taxon. When the issue of threatened and endangered species becomes important (see Chapter 2), then an identification to the species level is necessary.

Specialized Fields of Biology

There is no consensus or official list of the specialized fields within biology. However, most professional biologists agree about the major fields, and it will serve as a useful reference to list and discuss some of these fields.

BOTANY

Botany is the study of *plants*. This science is one of the largest, major subdisciplines of biology (along with zoology; see Chapter 1). It is the study of all plants, including microscopic plants, macroscopic plants, and all terrestrial and aquatic plants. Botany is in turn broken down into subspecialties.

ZOOLOGY

Zoology is the study of *animals*. It includes all animals, in a parallel sense with the study of botany. Thus the study of all microscopic animals, macroscopic animals, terrestrial animals and aquatic animals falls under the specialty of zoology. As in the case of botany, the field of zoology has many subspecialties.

Biology can also be subdivided into other specialized fields which center on the structure and function of living organisms. Historically, these fields include some of the older specialties within biology.

TAXONOMY

Taxonomy is the science of *classification* of living things. As early as the classical era of ancient Greece and Rome, philosophers attempted to classify the multi-tudinous organisms on earth. Linnaeus (see Chapter 1) is considered the father of modern taxonomy. He established the *binomial system* of nomenclature which we still use today. In addition to the genus and species, taxonomy includes other taxonomic categories. These were outlined in Appendix 1. Taxonomy may be subdivided into *plant taxonomy* and *animal taxonomy*.

MORPHOLOGY

Morphology is the study of the general *structure* of living organisms. It is also an ancient specialty within the general field of biology. It concerns itself with general structural features such as appendages, symmetry, and proportions. Under the subspecialty of *plant morphology*, for example, the study of leaves and branches would be of special interest. In the case of *animal morphology*, the fins of fish and the plumage of birds would be of special interest.

ANATOMY

Closely related to morphology is the study of anatomy. Anatomy focuses on gross *anatomical structures* readily obvious to the human eye. Anatomy generally concerns itself with organs and tissues. In the case of *plant anatomy*, the structure of leaves is an area of interest. *Animal anatomy* would include the structure of muscles, nerves, the liver, and other anatomical features. A special case of animal anatomy, of course, is *human anatomy*, which forms part of the medical curriculum.

HISTOLOGY

Histology is the study of *microscopic structure* of
plants and animals. The study of vascular tissues in
plants would be included in the field of *plant histology*.
Animal histology would include the microscopic structure
of nerve fibers, muscle fibers, glands, skin, and blood
vessels. *Human histology* is studied at medical schools.

CYTOLOGY

Closely related to histology is the field of cytology,
the science of *living cells*. All living organisms, with
minor exceptions, are composed of fundamental units called
cells. Cytology is concerned with the structure and
function of these living cells and their subcellular components. In the past few decades, scientists have used refined *biophysical* and *biochemical* methods to study living
cells and their components. The developments and findings
in such studies are having profound implications in many
facets of our modern world. Cytology is also divided into
plant cytology and *animal cytology*. However, the general
tendency today is to specialize in *general cytology*, because the evidence points to remarkable similarities in the
molecular machinery of plant and animal cells.

PHYSIOLOGY

Physiology deals with how living things *function*. All
living things are composed of atoms and molecules arranged
in complex hierarchies of organization. The study of how
organisms carry on respiration, metabolism, movement, and
the elimination of waste substances falls within the specialty of physiology. Physiology has become a very large
field, and consequently there are many subspecialties.
These include *bacterial physiology, plant physiology, animal physiology, insect physiology*, and *human physiology*.
Two other specialties, *biochemistry* and *biophysics,* which
deal with chemical and physical phenomena in living things,
had their early roots in physiology. Another important

subspecialty in recent decades is the science of *environ-mental physiology,* which deals with the functions of living organisms in a variety of environmental conditions.

PHARMACOLOGY AND TOXICOLOGY

Pharmacology is the science of how *chemical substances* (e.g., drugs) *interact with living organisms.* Much of the knowledge in this field has come from the development of therapeutic agents for humans and domestic animals. Conse-quently, it is often considered one of the medical sciences. Toxicology is really a subspecialty of pharmacol-ogy, with an emphasis on the *noxious* or *harmful effects* of chemicals on living organisms. Thus, how chemicals such as insecticides may cause harm is one of the topics that falls within the field of toxicology. *Environmental toxi-cology* is a rapidly-developing and important field.

PATHOLOGY

Both natural and human-caused events may cause *harm and disease* to plants and animals. Pathology is the study of such disease states. There is much practical value in the science of pathology, since it has direct applications to agriculture, forestry, veterinary medicine, and human medi-cine. It is now well-recognized that many environmental factors may be responsible for pathologic conditions in plants and animals.

ECOLOGY

Ecology is the study of how living *(biotic)* and non-living *(abiotic)* components of a defined area *interact and function.* The focus of ecology is usually on a complex system, generally called an ecosystem, rather than on indi-vidual organisms. Thus the study of how the physical, chemical, and biological components in a lake interact and function would fall within the field of *aquatic ecology.*

The study of the dynamic interactions among biotic and abiotic elements in a bay or estuary would fall within the specialized field of *marine ecology*. Finally, the study of the complex processes and biological interactions within a forest would lie within the subspecialty of *terrestrial ecology*. The application of ecology to large engineering projects has undergone important developments in the past decade.

OTHER SPECIALTIES

Biology is often subdivided into other specialties which concern themselves with specific groups of organisms. The following are examples of some of these specialties:

Bacteriology	(study of bacteria)
Mycology	(study of fungi)
Phycology	(study of algae)
Entomology	(study of insects)
Ichthyology	(study of fishes)
Herpetology	(study of reptiles)
Ornithology	(study of birds)
Mammology	(study of mammals)

Specialists within the field of entomology, for example, are concerned with the entire biology of insects. Thus they study the taxonomy, morphology, and physiology of insects. Conversely, they are not concerned with the biology of other groups (for example, algae).

Other specialties focus on humans and their needs. These include *agriculture, agronomy, forestry, fish and wildlife management,* and of course, the *medical sciences*.

SPECIALTIES AS APPLIED TO BIOLOGICAL COMPLEXITY

It is useful at this point to examine how the various biological specialties may be applied to the different levels of biological complexity outlined in Chapter 1. Although there is no precise match, Table 9 illustrates some appropriate and typical relationships. One point is illustrated rather clearly with this table, and that is the

TABLE 9. *Biological Specialties Frequently Applied to the Study of Various Levels of Biological Complexity.*

Level of biological complexity	Appropriate biological specialties
Biosphere	Ecology
Biome	Ecology
Ecosystem	Ecology
Community	Ecology, botany, zoology
Populations	Ecology, botany, zoology; many other specialties such as phycology, entomology, ichthyology
Organisms	Many specialties such as ecology, taxonomy, morphology, physiology, toxicology, pathology
Cells and tissues	Anatomy, histology, physiology, pathology, cytology, biophysics
Subcellular components	Cytology, biochemistry, biophysics
Molecules	Biochemistry, biophysics

tendency for ecology to be concerned with the large and
more complex systems. In contrast, the specialties of
biochemistry and biophysics tend to concentrate on subcellular and molecular components of living systems.
Stated another way, ecology tends to be holistic and look
at living systems in an integrative manner, whereas biochemistry and biophysics are analytical and tend to reduce
biological complexity into simpler components.

SPECIALTIES APPLIED TO ENVIRONMENTAL STUDIES

Practically all of these specialized fields of biology
play some role in environmental affairs. However, because
of the usual scale of engineering projects, it is the biological specialties that deal with populations, communities
and ecosystems that are applied most frequently to such
projects. In the larger and more complex projects it is
not unusual for the project sponsor to specify the types of
specialists (e.g., fishery biologist, ornithologist, toxi-
cologist, plant pathologist) that should be included on the
project team.

Some Comments on Biological Studies

The manner in which biological studies are now being conducted for environmental projects receives considerable attention. In the early years after NEPA, the biological studies for most environmental projects left much to be desired. Eventually the state-of-the-art improved, and considerable progress has been made in standardizing the methods for conducting these biological studies.

There is no single compendium of biological methods that are used in environmental research projects, and the complexity of the field is such that a single, comprehensive treatise is unlikely. Consequently, we are likely to see the continual development of various monographs of value mainly to specialists. Moreover, as the regulatory picture changes, the strategy, design and methods used in biological studies also change.

The objective of this appendix is to summarize certain aspects of biological studies that have practical relevance to engineering projects. The comments focus on major issues that are always relevant, irrespective of technical developments and changes in guidelines. In particular, these comments have practical significance to biologists and engineers who work together in the larger, more complex projects.

SPATIAL CONSIDERATIONS

Project areas that are spatially large and geographically complex offer considerable challenge. The spatial extent and typical inhomogeneity of habitat dictate a large number of sampling stations for quantitative studies or

surveys. An alternative is to make a general (non-quantitative) survey over an entire project area. It is usually
impractical to conduct extensive quantitative studies over
an entire project area. Various theories and practical
methods for field surveys have been developed. The ultimate objective in the selection of station locations is
to have accurate information at minimal cost. The selection of station locations is a major issue, and it must
be resolved in every project of any geographic scope.

STUDY PARAMETERS

Another important decision centers on the study parameters. For most environmental studies the objectives will
dictate the study parameters. The study parameters will
generally include the following parameter groups:

    abiotic factors in the aquatic environment
    aquatic flora
    aquatic fauna
    aquatic habitat
    abiotic factors in the terrestrial environment
    terrestrial flora
    terrestrial fauna
    terrestrial habitat

Each of these parameter groups can be divided into many
detailed parameters. It is important to select the parameters so that they are directly relevant to the overall
project. A common error is to include more parameters
than are really necessary to reach a sound decision.

TIMING CONSIDERATIONS

Ecosystems experience dynamic changes in both abiotic
and biotic components as the seasons change. Providing
there are no extreme time pressures on a project, it has
now become practically routine to include seasonal surveys.
In many cases the sampling frequency may be increased to
monthly or even biweekly surveys. This is especially true
for surveys of aquatic systems in the spring and summer

seasons. In other instances where population density does not change rapidly, a single survey may be appropriate. A good example of the latter case is seen in a survey of trees of major diameter (e.g., over 6 inches). Such a survey can usually be done at any time. However, a survey of under-story plants, especially for rare species, must be conducted at the appropriate time in the growing season. Again, these examples point out the importance of matching the timing for biological studies with the overall objectives of the project.

FIELD METHODS

Considerable effort has gone into the development of standard methods to conduct various field studies. In some cases the methods have been standardized by specialists in specific disciplines for many decades. Good examples include the methods used in limnology and fisheries biology in the aquatic environment. In the terrestrial environment, forestry methods and wildlife survey methods have also been developed for many years. The objective with most of these field methods is to sample biota and abiotic components so that quantitative analyses of the various parameters are possible. Some biological parameters lend themselves to quantitative methods; a good example is the analysis of phytoplankton density per liter. In other cases it is practically impossible to get an accurate measure of population density. This is especially true with highly mobile species such as some fish species and large wildlife species. In those instances the field surveys often include habitat surveys as one approach to estimate the probable carrying capacity. In general, habitat survey methods have not been standardized. Standard methods should be used whenever possible. In those cases where standard methods are not available, the field methods should be described in detail in the report.

LABORATORY METHODS

Similar comments can be made about laboratory methods used in biological studies. The pressures for standardization have increased over the past decade. There is every indication that progress in developing standard methods will continue. Government agencies have been particularly active in this area, as have private organizations. For example, the U.S. EPA and the American Society of Testing and Materials (ASTM) have active programs to develop standard methods in laboratory analyses. The amount of research and development activity centering on laboratory methods is likely to continue for some time. As with field methods, standard practices should be used for laboratory methods whenever possible.

STATISTICAL METHODS

Any studies that produce data-intensive results are always subject to scrutiny. The issues of accuracy and precision are always relevant to environmental studies. Such studies deal mostly with a non-uniform and dynamic environment. Therefore, study design, multiple samples, multiple analyses, and statistical analyses become important considerations. An important additional consideration, however, is the additional costs that are frequently necessary for the generation of multiple data bits. The project team must be alert, therefore, to the costs and benefits for doing extensive studies. The value of statistical methods is well recognized, and diligent efforts to produce reliable data are generally appropriate. However, there have been many instances of great statistical detail for some study components and only qualitative information for other, equally-relevant components. Thus a sense of balance and an understanding of the total objectives are important.

THE PERTINENT LITERATURE

An extensive literature is now available to guide
scientists and engineers in environmental studies. In
general, three types of documents are useful. The first
type of documentation includes all the various regulations
and guidelines published by the appropriate agencies.
These guidelines differ greatly in the amount of detail
directly relevant to biological studies. But in general
they contain much information which can assist in developing the overall conceptual design of the studies. In
particular, they give direction to the relative emphasis
for different environmental components. Another useful
source of information is the environmental assessment
literature that has accumulated for the past decade. For
example, over 10,000 environmental impact statements have
been prepared (1980). Many other environmental documents
of site-specific and region-specific interest are also
available. Not all of these documents are useful, of course;
however, these documents are a valuable resource on various biological studies, and every effort should be made to
consult them. The final source of information comes from
the standard literature on conducting biological studies.
Most biologists have some direct acquaintance with this
literature, both through their academic training and through
their subsequent work experience. (A number of these
standard references are included in the Selected References
listed in this book.)

Glossary

abiotic: non-living; characterized by the absence of
  life; in contrast with biotic.
abyssal: referring to the deep zone of the ocean.
acclimation: the adjustment of an organism to a new
  environmental state; commonly used to describe an
  adjustment to a temperature.
accumulation: the tendency of some molecules to build up
  in concentration; for example, the accumulation of
  pesticides in biota.
adaptation: the phenomenon in which an organism adjusts
  its processes to the ambient environmental conditions.
aerobic: pertaining to the presence of air.
agronomy: the biological discipline dealing with the
  study of field crop production and soil management.
algae: a large collection of simple plants, including
  microscopic and macroscopic species; algae do not have
  "true" roots, stems, flowers, etc.; (see also -
  plankton, periphyton, macrophyton).
allochthonous: organic material introduced into an
  aquatic ecosystem from external sources; for example,
  from trees along the shore.
Amphibia: a class of animals within the subphylum
  Vertebrata consisting of animals that live both on
  land and in water; for example, frog.
anadromous: term applied to fish that swim upstream to
  spawn; literally, going against the current; in
  contrast with catadromous; for example, salmon.
anaerobic: pertaining to the absence of air.
anatomy: the biological discipline dealing with the
  study of the structure of living organisms, especially larger species.
animal kingdom: that portion of the biosphere consisting
  of animals; all the animals; in contrast to the
  plant kingdom.

Annelida: a phylum of the animal kingdom consisting of segmented worms; for example, earthworm.

anoxic: pertaining to the absence of molecular oxygen.

aphotic: without light; relative absence of light, such as in deep water; in contrast with euphotic.

aquifer: a porous, water-bearing geologic formation.

Arthropoda: a phylum of the animal kingdom consisting of animals with jointed appendages; for example, insects, crustaceans.

association: the collection of species having a high probability of occurring together in the same local area.

autochthonous: organic material in an aquatic ecosystem generated within the system; for example, from aquatic plants.

autotroph: an organism that can manufacture its own food; literally, self-feeding; for example, plants; (see also - producer).

Aves: a class of animals within the subphylum *Vertebrata* consisting of animals with feathers, and which fly; for example, hawk.

bacteria: a group of microscopic organisms; generally important because of their functions as decomposers.

bacteriology: the biological discipline dealing with the study of bacteria and related organisms.

benthos: the organisms living at the bottom of an aquatic ecosystem; for example, benthic macroinvertebrate animals, as a portion of the total benthos; for example, clams, worms; the adjective form is "benthic".

bio: generalized prefix meaning life; for example, as in biology, the science of living things.

biochemistry: the biological discipline dealing with the study of chemical phenomena in living organisms.

biocide: generalized term for a chemical used to kill and control nuisance organisms; for example, insecticide; (see also - pesticide).

biogeochemical: pertaining to the cycling of atoms and molecules throughout the biosphere and non-living components of the earth.

biogeographic: pertaining to the geographic distribution and ranges of living organisms.

biomagnification: a phenomenon in which living organisms tend to build up concentrations of chemicals above the concentrations in the ambient environment.

biomass: the total amount of living material (mass) in a defined area or within a specific group; for example, the fish biomass of a lake; (see also - standing crop).

biome: the largest, generalized biological community; usually cover thousands of square miles on a subcontinental basis; for example, the North American tundra.

biophysics: the biological discipline dealing with the study of physical phenomena in living organisms.

biosphere: that portion of the earth where living organisms occur; a collective term for all living organisms.

biota: a collective term for all the plants (flora) and animals (fauna) of a given area.

biotic: pertaining to life or living organisms; in contrast with abiotic.

botany: the biological discipline dealing with the study of plants.

Brachiopoda: a phylum of the animal kingdom made up of relatively primitive organisms; for example, lamp shells.

brackish: referring to water which is a mixture of salt water and fresh water; generally found in coastal areas and estuaries.

Bryophyta: a phylum of the plant kingdom characterized by mosses, liverworts, and hornworts.

carcinogenic: literally cancer-causing; having the potential to cause cancer in living tissue; usually refers to chemicals.

carnivore: an animal (consumer) that requires another animal as a source of food; for example, mountain lion; (see also - secondary consumer).

carrying capacity: the amount of fauna (consumers) that a given habitat can sustain; generally based on an area of given dimensions.

catadromous: species of fish that descend rivers to spawn in the ocean; literally, going downstream; in contrast with anadromous; for example, eel.

cell: the basic structural unit of living organisms; cells carry on the basic life processes.

chaparral: a biome characterized by relatively dry climate and shrub-like vegetation.

Chlorophyta: a phylum of the plant kingdom consisting of the green algae.

Chordata: a phylum of the animal kingdom which includes the most advanced animals, including the vertebrates; for example, fish, birds, mammals.

Chrysophyta: a phylum of the plant kingdom consisting of yellow-green algae, golden algae, and diatoms.

class: a unit of taxonomy (or taxon) which subdivides a phylum into major groups; (see also - phylum).

climax stage: the final stage in the process of ecological succession; characterized by the climax vegetation and the climax community.

Coelenterata: a phylum of the animal kingdom consisting of primitive animals such as the hydras, jellyfishes, and corals.

community: a term within the structural hierarchy of ecology pertaining to many interacting populations in a given area.

compensation point: that depth in an aquatic system where the light conditions are such that the photosynthesis is just counterbalanced by the respiration, and thus there is no net production.

competition: interactions among biota in which they compete for food and habitat.

concentration-effect curve: the quantitative relationship between the concentration of a chemical and its effect on some living system; (see also - stress-effect curve).

coniferous: pertaining to conifers or evergreen trees; for example, pine trees, spruce.

consumer: an organism that utilizes existing organic matter as a source of food; (see also - heterotroph).

continental shelf: the relatively shallow, slowly-sloping extension of the continental land mass under the sea in the coastal area; generally extends out to 100 fathoms (200 meters).

continental slope: the bottom of the sea beyond the continental shelf, generally extending at a steeper gradient down into the bottom of the ocean.

critical habitat: a habitat with certain attributes which are critical for the survival of certain species; often applied to the habitat of threatened and endangered species.

Cyanophyta: a phylum of the plant kingdom consisting of the blue-green algae.

cytology: the biological discipline dealing with the study of the structure and function of living cells.

deciduous: pertaining to trees that shed their leaves on a seasonal basis; for example, oak trees, elm trees.

decomposer: an organism that breaks down complex organic matter into simpler constituents; for example, fungi, bacteria.

degradation: the breakdown of organic materials or organic
  molecules; usually occurs by various mechanisms;
  (see also - transformation).
density: an indication of the number of individuals of
  one or more species present in a given area; usually
  expressed as the number per unit area (or volume).
desert: a biome characterized by dry climate and vegetation adapted for relatively dry conditions; for example,
  cacti, yuccas.
detritivore: an organism that ingests detritus as its
  primary source of food; for example, earthworm.
detritus: the dead, organic material that accumulates in
  ecosystems, and provides energy through the organic
  molecules it contains.
detritus food chain: a food chain dependent upon detritus
  as its source of energy; in contrast with grazing food
  chain.
diagenesis: the chemical and physical transformations
  within sediments in aquatic ecosystems.
diversity: an indication of the variety of different
  species present in a given area; various quantitative
  measures of diversity have been developed.
Echinodermata: a phylum of the animal kingdom consisting
  of animals with spiny skins; for example, starfish,
  sea urchins.
ecological succession: the sequence of successional
  stages in a given area over a period of time,
  characterized by orderly changes in vegetation and
  associated biota.
ecology: the biological discipline dealing with the study
  of dynamic interrelationships among abiotic and biotic
  components of the environment.
ecosystem: an integrated, self-functioning system consisting of interactions among both abiotic and biotic
  elements; sizes of ecosystems may vary considerably.
ecotone: the interface between two types of ecosystems
  or communities; sometimes referred to as an edge
  effect.
ecotype: an organism of one species which is physically
  different from another organism of the same species
  living in a different geographic area.
Embryophyta: a subkingdom of the plant kingdom consisting
  of plants that form embryos; for example, mosses,
  ferns, flowers.

emergent vegetation: aquatic vegetation that emerges through the water surface; for example, cattails, pickerel weed.

endangered species: any species which is in danger of extinction throughout all or a significant portion of its range; (see also - threatened species).

entomology: the biological discipline dealing with the study of insects.

entrainment: a process in which small organisms (algae, plankton, larvae) are carried in the water as it flows through the condenser cooling water circuit.

epilimnion: the top layer of relatively warm and less-dense water in a lake or pond; in contrast with hypolimnion; the adjective form is "epilimnetic".

epiphyte: an organism that lives on the surface of a living plant.

estuary: a coastal area, generally semi-contained, in which fresh water intermixes with salt water.

Euglenophyta: a phylum of the plant kingdom consisting of the euglenoids.

Eumycophyta: a phylum of the plant kingdom consisting of the true fungi.

euphotic: the presence of light, generally sufficient for plant growth; in contrast with aphotic.

eutrophic: a condition in a water resource characterized by high levels of nutrients and relatively large biomass of plants; in contrast with an oligotrophic condition.

eutrophication: the process wherein the eutrophic condition progresses; commonly characterized by the evolution of a pond into a marsh or bog.

eutrophication, cultural: eutrophication which is accelerated by human activities.

eutrophication, natural: eutrophication which is not influenced by human activities.

evapotranspiration: the combined loss of moisture through the processes of evaporation and transpiration by plants.

family: a unit of taxonomy (or taxon) which subdivides an order into major groups; (see also - class and order).

fauna: biological term for animals.

fish control: a general term describing some of the management efforts in maintaining a desired fish population in a water resource; fish control often involves the use of biocides.

flora: biological term for plants.

flyway: a major geographic region of North America where migratory waterfowl tend to fly during their seasonal migrations.

food chain: a pathway of sequential steps involving consumers, representing different trophic levels; (see also - grazing food chain and detritus food chain).

food web: a complex network of various interrelationships among all food chains in an ecosystem.

forbs: a group of primary producers (plants) characterized by relatively short vegetation and flowering; does not include grasses and shrubs, but does include flowers and herbs.

fungi: a group of organisms generally classified as plants, but deriving their energy mainly as decomposers rather than as primary producers; (see also - decomposers).

genes: the chemical constituents in living cells which determine the inheritable characteristics of organisms.

genus: a unit of taxonomy (or taxon) which subdivides a family into major groups; (see also - order and family).

geographic range: the geographic area within which a population of a given species will live; the geographic range may be very extensive (for example, thousands of square miles).

grassland: a major biome characterized by light to moderate precipitation and grasses as the dominant vegetation.

graze: the living, plant material that is consumed by herbivores.

grazer: a consumer which depends upon primary producers (plants) as a source of food; (see also - herbivore, primary consumer).

grazing food chain: a food chain dependent upon primary producers (plants) as its source of energy; in contrast with detritus food chain.

habitat: the physical location or place where an organism lives, and where it derives food, shelter, and breeding area.

hectare: a unit of surface area which is equivalent to 2.47 acres (10,000 square meters); commonly used in ecological studies.

herbivore: a consumer (primary consumer) which depends upon primary producers (plants) as a source of food; (see also - grazer, primary consumer).

herpetology: the biological discipline dealing with the study of snakes and lizards.

heterotroph: an organism that requires other organisms or organic matter as sources of food; (see also - consumer).

histology: the biological discipline dealing with the study of plant and animal tissues.

home range: the geographic area within which an individual organism lives; varies considerably with the size of the organism; the term is generally applied to the larger animals; (see also - territory).

hypolimnion: the bottom layer of relatively cool, dense water in a lake or pond; in contrast with the epilimnion; the adjective form is "hypolimnetic".

ichthyology: the biological discipline dealing with the study of fish.

impact: a general term used in environmental assessment, meaning an effect or a consequence to a given action.

impingement: a process in which organisms such as fish are physically carried and held against the mesh of the intake screens of cooling water intake structures.

intertidal: an area of the coastal zone between the two tidal (high and low) levels; (see also - littoral).

invertebrates: animals without backbone; in contrast with vertebrates, or animals with backbone; for example, worms, clams, insects.

lacustrine: pertaining to an aquatic system that is impounded and does not flow; in contrast with riverine; (see also - lentic).

larvae: pre-adult forms of organisms; may refer to both invertebrates and vertebrate animals; for example, fish larvae.

legume: a type of plant characterized by the ability to enrich the soil through the presence of nitrogen-fixing bacteria in nodules in the roots; for example, peas, clover.

lentic: that type of aquatic system characterized by standing water; in contrast with lotic; (see also - lacustrine).

lichen: a relatively primitive plant characterized by a symbiotic relationship between an alga and a fungus.

litter: organic material, mostly composed of dead plant material; generally found on the forest floor.

littoral: pertaining to the shore and shallow water of a pond or coastal area; (see also - intertidal).

lotic: that type of aquatic system characterized by running water; in contrast with lentic; (see also - riverine).

macroalgae: that portion of the algal populations consisting of relatively large and visible species, (see also - macrophyton, algae).

macrophyton: literally large plants; a term commonly applied to aquatic plants that are readily visible; in contrast with phytoplankton; (also called macrophytes).

Mammalia: mammals, a class of animals within the subphylum *Vertebrata* consisting of animals that suckle their young; for example, dog.

mammology: the biological discipline dealing with the study of mammals.

menhaden: a species of marine fish; spends much of the juvenile stage in estuaries and coastal waters.

mesotrophic: a condition in a water resource characterized by nutrient levels and plant biomass which are intermediate to those found in eutrophic and oligotrophic systems; (see also - eutrophic, oligotrophic).

metabolism: biochemical processes in living organisms whereby molecules are both built up and broken down; involves such molecules as carbohydrates, fats, proteins and nucleic acids.

Metazoa: a subkingdom of the animal kingdom consisting of multicellular animals; in contrast with *Protozoa*.

migration: the movement of animal species from one location to another location; migrations may be daily and seasonal.

mitigation: an action taken to lessen the impact of another action.

molecule: a chemical substance made up of two or more atoms; for example, water, sugar.

Mollusca: a phylum of the animal kingdom consisting of soft-bodied animals; for example, clams, snails.

morphology: the biological discipline dealing with the study of general structure of living organisms.

moss: a plant that is a member of the phylum *Bryophyta*.

mutagenic: capable of inducing a genetic change.

mutualism, biological: a condition wherein growth and survival of two species are enhanced by an association.

mycology: the biological discipline dealing with the study of fungi.

Myxomycophyta: a phylum of the plant kingdom consisting of the slime molds.

natural selection: a process whereby the environmental conditions tend to select certain organisms for survival over long periods of time.

nekton: those organisms of the aquatic environment that are relatively strong swimmers, and hence do not float passively with currents; for example, fish; in contrast with plankton.

Nematoda: a phylum of the animal kingdom consisting of relatively primitive, worm-like animals; for example, roundworms.

neritic: a term referring to a zone of the marine environment as the water mass above the continental shelf.

niche: the role or function of an organism in an ecosystem; for example, primary producer, carnivore.

non-target species: organisms other than the target species intended for a biocide; (see also - target species).

nutrients: chemicals that are utilized by organisms in the synthesis of organic materials; frequently refers to such inorganic chemicals as phosphate and nitrate.

oceanic: a term referring to a zone of the marine environment beyond the continental shelf.

oligotrophic: a condition in a water resource characterized by low levels of nutrients and relatively small biomass of plants; in contrast with a eutrophic condition.

order: a unit of taxonomy (or taxon) which subdivides a class into major groups; (see also - phylum and class).

organism: a living individual; generalized term to describe a plant or animal.

ornithology: the biological discipline dealing with the study of birds.

overstory: the canopy of vegetation formed by the foliage of trees; in contrast with understory.

palustrine: an aquatic system characterized by no defined channel, no bedrock shoreline, and a generally well-developed aquatic vegetation; may include sections of rivers, lakes, or estuaries.

parasitism: an interaction between two species in which one population adversely affects the other by direct attack, but is also dependent on the other.

pathology: the biological discipline dealing with the study of diseased conditions in plants and animals.

pelagic: that portion of the marine environment consisting of the open sea.

periphyton: groups of organisms clinging and growing on stems, leaves, and other surfaces in the aquatic environment.

permafrost: permanently frozen deeper layers under the
arctic tundra; (see also - tundra).

persistence: the property of a chemical (e.g., a pesticide)
whereby it tends to persist in the environment in
its original molecular form; in contrast with
degradation.

pesticide: a chemical substance used to kill and/or
control nuisance species; for example, insecticide;
(see also - biocide).

Phaeophyta: a phylum of the plant kingdom consisting of
brown algae.

pharmacology: the biological discipline dealing with the
study of how chemical substances may affect living
organisms.

photosynthesis: the synthesis of carbohydrates from
inorganic chemicals using light as an energy source.

phycology: the biological discipline dealing with the
study of algae.

phylum: the largest taxonomic unit which subdivides the
biota into major groups; the term applies to both
the plant and animal kingdoms; (see also - class,
order, family, genus, and species).

physiology: the biological discipline dealing with the
study of how living organisms function.

phytoplankton: that portion of the plankton consisting
of microscopic plants; in contrast with macrophyton;
(see also - plankton).

pioneer species: plant species that appear early in the
successional sequence and colonize a bare rock or
cleared area.

Pisces: a class of animals within the subphylum *Vertebrata*
consisting of the fishes; for example, sharks, trout.

plankton: organisms floating passively in the water and
carried by water currents; in contrast with nekton;
(see also - phytoplankton, zooplankton).

plant association: a community of different plant species,
usually quite characteristic of the regional
environment.

plant kingdom: that portion of the biosphere consisting
of plants; all of the plants; in contrast with the
animal kingdom.

Platyhelminthes: a phylum of the animal kingdom consisting
of the flatworms.

population: a group of individuals of the same species.

population dynamics: the changes in populations of biota
brought about by various ecological processes.

Porifera: a phylum of the animal kingdom consisting of sponges.

prairie: the grassland biome of North America; characterized by relatively light to moderate rainfall; (see also - grassland).

predator: a carnivore that preys upon other animals for a food source; (see also - prey).

prey: the animals that are preyed upon by predator species; (see also - predator).

primary consumer: the first trophic level of consumers in the food chain; (see also - herbivore, grazer).

primary producer: see - producer.

producer: an organism that can manufacture food from inorganic nutrients; includes plants; (see also - autotroph).

productivity: a measure of the amount of energy (or biomass) produced by a biotic group in a specific time period; (see also - producer).

Protozoa: a subkingdom and phylum of the animal kingdom consisting of unicellular animals; in contrast with *Metazoa;* for example, amoeba.

pyramid of biomass: a relationship in which the biomass in progressively higher trophic levels becomes smaller; (see also - pyramid of numbers).

pyramid of numbers: a relationship in which the numbers of organisms in progressively higher trophic levels become smaller; (see also - pyramid of biomass).

Pyrrophyta: a phylum of the plant kingdom consisting of dinoflagellates and cryptomonads.

Reptilia: a class of animals within the subphylum *Vertebrata* consisting of reptiles, lizards and snakes.

respiration: the biological process of taking in oxygen and giving off carbon dioxide.

Rhodophyta: a phylum of the plant kingdom consisting of the red algae.

riffle: a stretch of stream characterized by turbulent water, usually flowing along a gradient over rocks.

riparian: pertaining to the banks of a river; commonly applied to biota along the banks.

riverine: pertaining to an aquatic system that is flowing along a gradient; in contrast with lacustrine; (see also - lotic).

roadkill: the loss of animals through collisions with motor vehicles.

Rotifera: a phylum of the animal kingdom consisting of relatively small and primitive animals called rotifers.

salinity: a measure of the amount of dissolved solids in water, such as sea water or brackish water.

Schizomycophyta: the phylum of the plant kingdom consisting of the bacteria.

secondary consumer: a consumer that feeds upon primary consumers; (see also - carnivore).

sedge: types of plants characterized by grasslike or rushlike appearance; commonly found in marshy areas.

sediment: materials, usually of geologic origin, which are carried by water and which eventually separate out to form layers on the bottom.

species: the lowest unit of taxonomy (or taxon); it subdivides a genus into the scientific names of individual organisms that are genetically distinct; (see also - family and genus).

stability: a measure of the ability of an ecosystem to maintain itself in or near a natural equilibrium state in the face of some stress or disturbance.

standing crop: the amount of living material (or biomass) present at any one time for some biotic group; (see also - biomass).

stocking: the introduction of a species (usually fish species) as part of a program of managing fishery resources.

stratification: a process whereby the upper, warmer layers of water form a stable, stratified system above the lower, cooler water in a lake; (see also - epilimnion, hypolimnion, thermocline).

stress: any physical, chemical or biological processes (or combination of processes) which tend to affect biota such that they are displaced from an existing optimum condition.

stress-effect curve: the quantitative relationship between the magnitude of a stress and its effect on some living system; (see also - concentration-effect curve).

substrate: the immediate physical material on or in which an organism lives; for example, soil.

succession: see - ecological succession.

taiga: another name for the northern coniferous forest; characterized by evergreen trees such as pine and hemlock.

target species: the species intended to be controlled or killed by a biocide; (see also - non-target species).

taxon: a unit of biological classification; (see also - phylum, class, order, family, genus and species).

taxonomy: the biological discipline dealing with the study of the classification of plants and animals.

teratogenic: having the potential to cause deformations in living organisms; usually refers to chemical compounds.

territory: a circumscribed area, generally within a home range, which an animal will actively defend against competitors; (see also - home range).

Thallophyta: a subkingdom of the plant kingdom consisting of plants without roots, stems and leaves; for example, algae.

thermal stratification: see - stratification.

thermocline: the zone between the upper, warmer water and the lower, colder water of an aquatic system; (see also - epilimnion, hypolimnion).

threatened species: any species which is likely to become an endangered species within the foreseeable future throughout all or a significant portion of its range; (see also- endangered species).

tissue: a collection of living cells structurally similar and integrated to perform a similar function; for example, muscle tissue.

toxicity: a quantitative assessment of the ability of a compound to kill or cause harm; may include acute toxicity (short-term exposure or action) and chronic toxicity (longer-term).

toxicology: the biological discipline dealing with the study of the harmful effects of chemicals on living organisms.

Tracheophyta: a phylum of the plant kingdom consisting of plants with vascular tissues; for example, trees.

transformation: the chemical changes that substances may be subjected to in the environment; (see also - degradation).

trophic: the condition of a pond or lake relevant to nutrient levels and plant biomass; (see also - oligotrophic and eutrophic); also, the generalized concept of the relationships among biota in a food web.

trophic level: the biological energy transfer level; the position in a trophic pyramid or food chain.

trophic magnification: the process whereby potentially toxic chemicals are concentrated as they pass through the various trophic levels of a food chain.

trophic state: the state or condition of a lake or pond
     with respect to nutrient levels and plant biomass;
     (see also - eutrophic, mesotrophic, oligotrophic).
tundra: a vast biome along the northern part of North
     America characterized by low average temperatures,
     short growing season and a mat-like vegetation.
understory: the bed of vegetation formed by the foliage
     of herbs and grasses; in contrast with overstory.
upwelling: the phenomenon in an aquatic environment
     whereby vertical currents bring up water from
     lower depths; these waters are usually rich in
     nutrients.
Vertebrata: a subphylum of the phylum *Chordata* character-
     ized by animals with a backbone (vertebrae); includes
     fish, birds, mammals; (see also - *Chordata*).
vertebrates: animals with backbone; in contrast with
     invertebrates, or animals without backbone; for
     example, fish, birds, mammals.
wildlife: a general term referring to animals not
     considered domestic species; in its broad meaning
     it includes both fish and terrestrial species.
windthrow: the action of wind against trees, especially
     at an edge or ecotone where the trees are more
     vulnerable.
yield: pertaining to the productivity of an ecosystem;
     generally specified in reference to some group, such
     as fish yield.
zoology: the biological discipline dealing with the
     study of animals.
zooplankton: that portion of the plankton consisting of
     animal species; (see also - plankton).

Selected References

This section contains selected references of potential value to engineers concerned with environmental biology. The list also includes the references cited in the text. No attempt is made to include the conventional references in environmental engineering and the voluminous literature in environmental regulations.

GENERAL ECOLOGY AND ENVIRONMENTAL SCIENCE

Annual Reviews Inc. 1970- . *Annual Review of Ecology and Systematics*. 4139 El Camino Way, Palo Alto, California 94306.

Bailey, R.G. 1978. *Descriptions of the Ecoregions of the United States*. Forest Service. U.S. Department of Agriculture, Ogden, Utah.

Clarke, G.L. 1965. *Elements of Ecology*. Revised Edition, John Wiley & Sons, New York, New York.

DeSanto, R.S. 1978. *Concepts of Applied Ecology*. Springer-Verlag, New York, New York.

Hinckley, A.D. 1976. *Applied Ecology: A Nontechnical Approach*. Macmillan, New York, New York.

Kormondy, E.J. 1969. *Concepts of Ecology*. Prentice-Hall, Englewood Cliffs, New Jersey.

Krebs, C.J. 1972. *Ecology: The Experimental Analysis of Distribution and Abundance*. Harper and Row, New York, New York.

McNaughton, S.J. and L.L. Wolf. 1973. *General Ecology.*
        Holt, Rinehart and Winston, New York, New York.

Odum, E.P. 1971. *Fundamentals of Ecology.* Third Edition,
        Saunders, Philadelphia, Pennsylvania.

Odum, H.T. 1971. *Environment, Power and Society.* John
        Wiley & Sons, New York, New York.

Pielou, E.C. 1975. *Ecological Diversity.* Wiley-
        Interscience, Somerset, New Jersey.

Pielou, E.C. 1977. *Mathematical Ecology.* Wiley-
        Interscience, Somerset, New Jersey.

Ricklefs, R.E. 1973. *Ecology.* Chiron Press, Newton,
        Massachusetts.

Shelford, V.E. 1963. *The Ecology of North America.*
        University of Illinois Press, Urbana, Illinois.

Smith, R.L. 1974. *Ecology and Field Biology.* Second
        Edition, Harper and Row, New York, New York.

Southwick, C.H. 1976. *Ecology and the Quality of Our
        Environment.* Second Edition, D. Van Nostrand
        Company, New York, New York.

U.S. Environmental Protection Agency, (EPA). 1973.
        Common Environmental Terms: A Glossary.
        Washington, D.C. 20460.

Van Dyne, G.M., (Editor). 1969. *The Ecosystem Concept
        of Natural Resource Management.* Academic Press,
        New York, New York.

Watt, K.E.F. 1966. *Systems Analysis in Ecology.*
        Academic Press, New York, New York.

Watt, K.E.F. 1973. *Principles of Environmental Science.*
        McGraw-Hill Book Company, New York, New York.

TERRESTRIAL ENVIRONMENT

Brockman, C.F. 1968. *Trees of North America.* Golden
     Press, New York, New York.

Cloudsley - Thompson, J.L. 1975. *Terrestrial Environ-
     ments.* Halsted Press, John Wiley and Sons, New
     York, New York.

Daubenmire, R.F. 1974. *Plants and Environment: A Text-
     book of Plant Autecology.* Third Edition, John
     Wiley & Sons, New York, New York.

Fernald, M.L. 1950. *Gray's Manual of Botany.* Eighth
     Edition, Corrected Printing, 1970, Van Nostrand
     Reinhold Company, New York, New York.

Hitchcock, A.S. 1971. *Manual of Grasses of the United
     States.* Vol. I, Dover Publications, New York,
     New York.

Hitchcock, A.S. 1971. *Manual of Grasses of the United
     States.* Vol. II, Dover Publications, New York,
     New York.

Little, E.L., Jr. 1953. *Check List of Native and
     Naturalized Trees of the United States (Including
     Alaska).* Agriculture Handbook No. 41, Forest
     Service, U.S. Department of Agriculture,
     Washington, D.C.

Little, E.L., Jr. 1971. *Atlas of United States Trees:
     Vol. 1, Conifers and Important Hardwoods.*
     Miscellaneous Publication No. 1146, Forest Service,
     U.S. Department of Agriculture, Washington, D.C.

Litton, R.B., Jr. 1968. *Forest Landscape, Description
     and Inventories, A Basis for Land Planning
     and Design.* U.S. Department of Agriculture,
     Forest Service Research Paper PSW-49, Berkeley,
     California.

Mueller-Dombois, D. and H. Ellenberg. 1974. *Aims and
     Methods of Vegetation Ecology.* John Wiley & Sons,
     New York, New York.

Oosting, H.J.   1956.   *The Study of Plant Communities.*
      Second Edition, W.H. Freeman and Company, San
      Francisco, California.

Peterson, R.T. and M. McKenny.   1968.   *A Field Guide
      to Wildflowers of Northeastern and North-Central
      North America.*   Houghton Mifflin Company,
      Boston, Massachusetts.

Petrides, G.A.   1972.   *A Field Guide to Trees and Shrubs.*
      Second Edition, Houghton Mifflin Company, Boston,
      Massachusetts.

Platt, R.   1965.   *The Great American Forest.*   Prentice-
      Hall, Englewood Cliffs, New Jersey.

Reichle, D.E., (Editor).   1970.   *Analysis of Temperate
      Forest Ecosystems.*   Springer-Verlag, New York,
      New York.

Spurr, S.H. and B.V. Barnes.   1973.   *Forest Ecology.*
      Second Edition, Ronald Press Company, New York,
      New York.

Swan, L.A. and C.S. Papp.   1972.   *The Common Insects of
      North America.*   Harper and Row, New York, New York.

Symonds, G.W.D.   1963.   *The Shrub Identification Book.*
      William Morrow and Company, New York, New York.

FRESHWATER ENVIRONMENT

Bennett, G.W.   1971.   *Management of Lakes and Ponds.*
      Second Edition, Van Nostrand Reinhold, New York,
      New York.

Canale, R.P.   1976.   *Modeling Biochemical Processes in
      Aquatic Ecosystems.*   Ann Arbor Science Pub-
      lishers, Ann Arbor, Michigan.

Cole, G.A.   1979.   *Textbook of Limnology.*   Second
      Edition, Mosby, St. Louis, Missouri.

Edmondson, W.T., (Editor). 1959. *Fresh-Water Biology.* Second Edition, John Wiley & Sons, New York, New York.

Frey, D.G., (Editor). 1966. *Limnology in North America.* University of Wisconsin Press, Madison, Wisconsin.

Hutchinson, G.E. 1957. *A Treatise on Limnology: Volume I, Geography, Physics, and Chemistry.* John Wiley & Sons, New York, New York.

Hutchinson, G.E. 1967. *A Treatise on Limnology: Volume II, Introduction to Lake Biology and the Limno-plankton.* John Wiley & Sons, New York, New York.

Hutchinson, G.E. 1975. *A Treatise on Limnology: Volume III, Limnological Botany.* John Wiley & Sons, New York, New York.

Hynes, H.B.N. 1970. *The Ecology of Running Waters.* University of Toronto Press, Toronto, Ontario.

Lind, O.T. 1979. *Handbook of Common Methods in Limnology.* Second Edition, Mosby, St. Louis, Missouri.

Litton, R.B., Jr., R.J. Tetlow, J. Sorensen and R.A. Beatty. 1974. *Water and Landscape: An Aesthetic Overview of the Role of Water in the Landscape.* Water Information Center, Inc., 44 Sintsink Drive East, Port Washington, New York 11050.

Macan, T.T. 1974. *Freshwater Ecology.* Second Edition, John Wiley & Sons, New York, New York.

National Academy of Sciences. 1969. *Eutrophication: Causes, Consequences, Correctives.* Washington, D.C.

National Water Commission. 1973. *Water Policies for the Future.* Water Information Center, Inc., 44 Sintsink Drive East, Port Washington, New York 11050.

Oglesby, R.T., C.A. Carlson, and J.A. McCann. 1972.
       River Ecology and Man. Academic Press, New York,
       New York.

Pennak, R.W. 1978. *Fresh-Water Invertebrates of the
       United States.* Second Edition, Wiley-Interscience,
       Somerset, New Jersey.

Prescott, G.W. 1970. *How to Know the Freshwater Algae.*
       Second Edition, W.C. Brown, Dubuque, Iowa.

Ruttner, F. 1963. *Fundamentals of Limnology.* Third
       Edition, University of Toronto Press, Toronto,
       Ontario.

U.S. Environmental Protection Agency, (EPA). March 1975.
       Plankton Analysis Training Manual. Office of
       Water Programs Operations, National Training
       Center, Cincinnati, Ohio 45268. (EPA-430/1-75-004).

U.S. Environmental Protection Agency, (EPA). April 1975.
       *Freshwater Biology and Pollution Ecology, Training
       Manual.* Office of Water Programs Operations.
       National Training Center, Cincinnati, Ohio
       45268. (EPA-430/1-75-005).

Usinger, R.L., (Editor). 1971. *Aquatic Insects of
       California with Keys to North American Genera and
       California Species.* University of California
       Press, Berkeley, California.

Weber, C.I., (Editor). 1973. *Biological Field and
       Laboratory Methods for Measuring the Quality of
       Surface Waters and Effluents.* National Environ-
       mental Research Center, Office of Research and
       Development, U.S. Environmental Protection
       Agency, Cincinnati, Ohio 45268. (EPA-670/4-73-001).

Welch, P.S. 1948. *Limnological Methods.* McGraw-Hill,
       New York, New York.

Welch, P.S. 1952. *Limnology.* Second Edition, McGraw-
       Hill, New York, New York.

Wetzel, R.G. 1975. *Limnology.* W.B. Saunders, Phila-
       delphia, Pennsylvania.

Wood, R.D. 1975. *Hydrobotanical Methods*. University
    Park Press, Baltimore, Maryland.

MARINE AND ESTUARINE ENVIRONMENT

Abbott, R.T. 1961. *How to Know the American Marine
    Shells*. Signet Key Books, New York, New York.

Clark, J.R. 1977. *Coastal Ecosystem Management: A
    Technical Manual for the Conservation of Coastal
    Zone Resources*. Wiley - Interscience, Somerset,
    New Jersey.

Dawson, E.Y. 1956. *How to Know the Seaweeds*. W.C.
    Brown, Dubuque, Iowa.

Friedrich, H.N. 1970. *Marine Biology*. University of
    Washington Press, Seattle, Washington.

Gosner, K.L. 1971. *Guide to the Identification of
    Marine and Estuarine Invertebrates: Cape
    Hatteras to the Bay of Fundy*. Wiley-Interscience,
    Somerset, New Jersey.

Green, J. 1968. *The Biology of Estuarine Animals*.
    University of Washington Press, Seattle,
    Washington.

Lauff, G.H., (Editor). 1967. *Estuaries*. American
    Association for the Advancement of Science, 1515
    Massachusetts Avenue, N.W., Washington, D.C.
    20005.

McConnaughey, B.H. 1978. *Introduction to Marine Biology*.
    Third Edition, Mosby, St. Louis, Missouri.

Meadows, P.S. and J.I. Campbell. 1978. *An Introduction
    to Marine Science*. Wiley-Interscience, Somerset,
    New Jersey.

Menzies, R.J., R.Y. George and G.T. Rowe. 1973.
    *Abyssal Environment and Ecology of the World
    Oceans*. Wiley-Interscience, Somerset, New
    Jersey.

Miner, R.W. 1950. *Field Book of Seashore Life*. Putnam,
    New York, New York.

Morris, P.A. 1951. *A Field Guide to the Shells of Our
    Atlantic and Gulf Coasts*. Second Edition,
    Houghton Mifflin Company, Boston, Massachusetts.

Odum, H.T., B.J. Copeland, and E.A. McMahan, (Editors).
    1974. *Coastal Ecological Systems of the United
    States*. Volumes I, II, III and IV. The Conser-
    vation Foundation, Washington, D.C. 20036.

Officer, C.B. 1976. *Physical Oceanography of Estuaries
    (And Associated Coastal Waters)*. Wiley-Inter-
    science, Somerset, New Jersey.

Reimold, R.J. and W.H. Queen, (Editors). 1974. *Ecology
    of Halophytes*. Academic Press, New York, New
    York.

Ricketts, E.F. and J. Calvin. 1968. *Between Pacific
    Tides*. Fourth Edition, Stanford University
    Press, Palo Alto, California.

Schlieper, C., (Editor). 1972. *Research Methods in Marine
    Biology*. University of Washington Press, Seattle,
    Washington.

Sverdrup, H.U., M.W. Johnson, and R.H. Fleming. 1942.
    *The Oceans: Their Physics, Chemistry and
    General Biology*. Prentice-Hall, Englewood
    Cliffs, New Jersey.

Tait, R.V. and R.S. DeSanto. 1974. *Elements of Marine
    Ecology, An Introductory Course*. Springer-
    Verlag, New York, New York.

U.S. Environmental Protection Agency, (EPA). 1974.
    *Proceedings of Seminar on Methodology for
    Monitoring the Marine Environment*. Office of
    Research and Development, Washington, D.C. (EPA-
    660/4-74-004).

Weyl, P.K. 1970. *Oceanography: An Introduction to the
    Marine Environment*. John Wiley & Sons, New York,
    New York.

Zottoli, R.   1978.   *Introduction to Marine Environments.*
        Second Edition, Mosby, St. Louis, Missouri.

WETLANDS

Cowardin, L.M., V. Carter, F.C. Golet, and E.T. LaRoe.
        December 1979.   *Classification of Wetlands and
        Deepwater Habitats of the United States.*   Fish
        and Wildlife Services, U.S. Department of the
        Interior, Washington, D.C. 20240. (FWS/OBS-79/31).

Darnell, R.W., W.E. Pequegnat, B.M. James, F.J. Benson,
        and R.A. Defenbaugh.   1976.   *Impacts of Construc-
        tion Activities in Wetlands of the United States.*
        Corvallis Environmental Research Laboratory,
        Office of Research and Development, U.S. Environ-
        mental Protection Agency, Corvallis, Oregon
        97330. (EPA 600/3-76-045).

Fassett, N.C.   1957.   *A Manual of Aquatic Plants.*
        University of Wisconsin Press, Madison, Wisconsin.

Greeson, P.E., J.R. Clark, and J.E. Clark, (Editors).
        1978.   *Wetland Functions and Values: The State
        of Our Understanding.*   American Water Resources
        Association, Minneapolis, Minnesota.

Garbisch, E.W., Jr.   1977.   *Recent and Planned Marsh
        Establishment Work Throughout the Contiguous
        United States: A Survey and Basic Guidelines.*
        Office, Chief of Engineers, U.S. Army,
        Washington, D.C. 20314.

Good, R.E., D.F. Whigham and R.L. Simpson, (Editors).
        1978.   *Freshwater Wetlands: Ecological Processes
        and Management Potential.*   Academic Press, New
        York, New York.

Horwitz, E.L.   1978.   *Our Nation's Wetlands: An
        Interagency Task Force Report.*   U.S. Government
        Printing Office, Washington, D.C. 20402.

Hotchkiss, N.   1972.   *Common Marsh, Underwater & Floating-
        Leaved Plants of the United States and Canada.*
        Dover Publications, New York, New York.

Lynch, M.P., B.L. Laird, N.B. Theberge, and J.C. Jones, (Editors). 1976. *An Assessment of Estuarine and Nearshore Marine Environments.* Office of Biological Services, Fish and Wildlife Service, U.S. Department of the Interior, Washington, D.C.

Odum, H.T., B.J. Copeland and E.A. McMahan, (Editors). 1974. *Coastal Ecological Systems of the United States.* (Four Volumes), The Conservation Foundation, Washington, D.C. 20036.

Prescott, G.W. 1969. *How to Know the Aquatic Plants.* W.C. Brown, Dubuque, Iowa.

Reppert, R.T., W. Sigleo, E. Stakhiv, L. Messman, and C. Meyers. 1979. *Wetland Values: Concepts and Methods for Wetlands Evaluation.* U.S. Army Corps of Engineers, Institute for Water Resources, Fort Belvoir, Virginia 22060.

Sather, J.H., (Editor). July 1976. *Proceedings of the National Wetland Classification and Inventory Workshop, 1975.* Office of Biological Services, Fish and Wildlife Service, U.S. Department of the Interior, Washington, D.C. 20240. (FWS/OBS - 76/09).

Schamberger, M., C. Short, and A. Farmer. 1978. *Evaluating Wetlands as Wildlife Habitat: Proceedings, National Symposium on Wetlands.* Division of Ecological Services, U.S. Fish and Wildlife Service, Fort Collins, Colorado 80526.

Shaw, S.P. and C.G. Fredine. 1956. *Wetlands of the United States.* Fish and Wildlife Service, Circular 39, U.S. Department of the Interior, Washington, D.C. 20240.

Teal, J.M. and M. Teal. 1969. *Life and Death of the Salt Marsh.* Atlantic, Little Brown and Company, Boston, Massachusetts.

Tesky, R.O. and T.M. Hinckley. 1977. *Impact of Water Level Changes on Woody Riparian and Wetland Communities.* (Three Volumes), Office of Biological

Services, Fish and Wildlife Service, U.S. Department of the Interior, Columbia, Missouri 65201.

The White House. May 24, 1977. Executive Order 11990, Protection of Wetlands. *Federal Register,* Vol. 42, No. 101 - Wednesday, May 25, 1977, 26961-26965.

FISH AND WILDLIFE

American Fisheries Society. 1970. *A List of Common and Scientific Names of Fishes from the United States and Canada.* Special Publication No. 6, Third Edition, Washington, D.C.

Brokaw, H.P., (Editor). 1978. *Wildlife and America.* U.S. Government Printing Office, Washington, D.C. 20402.

Burt,W.H. and R.P. Grossenheider. 1964. *A Field Guide to the Mammals.* Houghton Mifflin Company, Boston, Massachusetts.

Carlander, K.D. 1969. *Handbook of Freshwater Fishery Biology.* Vol. 1, Iowa State University Press, Ames, Iowa.

Carlander, K.D. 1977. *Handbook of Freshwater Fishery Biology.* Vol. 2, Iowa State University Press, Ames, Iowa.

Conant, R. 1958. *A Field Guide to Reptiles and Amphibians of Eastern North America.* Houghton Mifflin Company, Boston, Massachusetts.

Eddy, S. 1969. *How to Know the Freshwater Fishes.* Second Edition, W.C. Brown, Dubuque, Iowa.

Eddy, S. and T. Surber. 1960. *Northern Fishes: With Special Reference to the Upper Mississippi Valley.* Revised Edition, C.T. Branford, Newton Centre, Massachusetts.

Gerking, S.D.   1978. *Ecology of Freshwater Fish Produc-
        tion.* Halsted Press, John Wiley & Sons, New York,
        New York.

Giles, R.H., (Editor).   1971. *Wildlife Management
        Techniques.* Third Edition, Revised, The Wildlife
        Society, Washington, D.C.

Gulland, J.A.   1974. *The Management of Marine Fisheries.*
        University of Washington Press, Seattle,
        Washington.

Hall, G.E., (Editor).   1971. *Reservoir Fisheries and
        Limnology.* Special Publication No. 8, American
        Fisheries Society, Washington, D.C.

Headstrom, R.   1970. *A Complete Field Guide to Nests
        in the United States, Including Those of Birds,
        Mammals, Insects, Fishes, Reptiles, and Amphibians.*
        Ives Washburn, New York, New York.

Hubbs, C.L. and K.F. Lagler.   1958. *Fishes of the Great
        Lakes Region.* University of Michigan Press,
        Ann Arbor, Michigan.

Jordan, D.S. and B.W. Evermann.   1969. *American Food
        and Game Fishes: A Popular Account of All
        Species Found in America North of the Equator,
        with Keys for Ready Identification, Life
        Histories and Methods of Capture.* Dover
        Publications, New York, New York.

Lagler, K.F., 1956. *Freshwater Fishery Biology.*
        Second Edition, W.C. Brown, Dubuque, Iowa.

Lagler, K.F., J.E. Bardach, R.R. Miller, and D.R.M.
        Passino. 1977. *Ichthyology.* Second Edition,
        John Wiley & Sons, New York, New York.

Martin, A.C., H.S. Zim, and A.L. Nelson.   1951. *American
        Wildlife and Plants: A Guide to Wildlife Food
        Habits.* Dover Publications, New York, New York.

McElroy, T.P., Jr.   1974. *The Habitat Guide to Birding.*
        Alfred A. Knopf, New York, New York.

Palmer, E.L. and H.S. Fowler. 1975. *Fieldbook of Natural History.* Second Edition, McGraw-Hill Book Company, New York, New York.

Palmer, R.S. 1954. *The Mammal Guide.* Doubleday, Garden City, New York.

Robbins, C.S., B. Bruun, and H.S. Zim. 1966. *Birds of North America.* Golden Press, New York, New York.

Scott, W.B. and E.J. Crossman. 1973. *Freshwater Fishes of Canada.* Department of the Environment, Ottawa.

U.S. Department of the Interior, Fish and Wildlife Service. 1979. List of Endangered and Threatened Wildlife and Plants, Republication. *Federal Register,* Vol. 44, No. 12, Wednesday, January 17, 1979, 3636-3654.

ENVIRONMENTAL STRESS AND TOXICOLOGY

Ariens, E.J., A.M. Simonis and J. Offermier. 1976. *Introduction to General Toxicology.* Academic Press, New York, New York.

Butler, G.C., (Editor). 1978. *Principles of Ecotoxicology.* Wiley-Interscience, Somerset, New Jersey. (Scope Report 12).

Eisenbud, M. 1973. *Environmental Radioactivity.* Second Edition, Academic Press, New York, New York.

Glass, D.C. and J.E. Singer. 1972. *Urban Stress: Experiments on Noise and Social Stressors.* Academic Press, New York, New York.

Goldberg, E.D. 1972. *A Guide to Marine Pollution.* Gordon and Breach, New York, New York.

Haque, R., (Editor). 1979. *Dynamics, Exposure and Hazard Assessment of Toxic Chemicals in the Environment.* Ann Arbor Science Publishers, Ann Arbor, Michigan.

Hart, C.W., Jr. and S.L.H. Fuller, (Editors). 1974.
Pollution Ecology of Freshwater Invertebrates.
Academic Press, New York, New York.

Khan, M.A.Q. and J.P. Bederka, Jr., (Editors). 1974.
Survival in Toxic Environments. Academic Press,
New York, New York.

Kraybill, H.F. and M.A. Mehlman, (Editors). 1977.
Environmental Cancer. Halsted Press, John
Wiley & Sons, New York, New York.

Lockwood, A.P.M., (Editor). 1976. *Effects of Pollutants
on Aquatic Organisms.* Cambridge University
Press, New York, New York.

Marking, L.L. and R.S. Kimerle, (Editors). May 1979.
Aquatic Toxicology. ASTM STP 667, American
Society for Testing and Materials (ASTM),
Philadelphia, Pennsylvania.

Matsumura, F., G.M. Boush and T. Misato, (Editors).
1972. *Environmental Toxicology of Pesticides.*
Academic Press, New York, New York.

Miller, M.W. and J.N. Stannard, (Editors). 1976.
*Environmental Toxicity of Aquatic Radionuclides:
Models and Mechanisms.* Ann Arbor Science
Publishers, Ann Arbor, Michigan.

Mitchell, R., (Editor). 1972. *Water Pollution
Microbiology.* Wiley-Interscience, Somerset, New
Jersey.

Mitchell, R., (Editor). 1978. *Water Pollution Microbiology, Volume 2.* Wiley-Interscience, Somerset,
New Jersey.

Mudd, J.B. and T.T. Kozlowski, (Editors). 1975.
Responses of Plants to Air Pollution.
Academic Press, New York, New York.

Mussell, H. and R.C. Staples, (Editors). 1979. *Stress
Physiology in Crop Plants.* Wiley-Interscience,
Somerset, New Jersey.

Naegele, J.A., (Editor). 1973. *Air Pollution Damage to Vegetation*. Advances in Chemistry Series, 122, American Chemical Society, Washington, D.C.

O'Brien, R.D. and I. Yamamoto, (Editors). 1970. *Biochemical Toxicology of Insecticides*. Academic Press, New York, New York.

Rudd, R. L. 1977. *Environmental Toxicology: A Guide to Information Sources*. Gale Research Company, Book Tower, Detroit, Michigan.

Schubel, J.R. and B.C. Marcy, Jr., (Editors). 1978. *Power Plant Entrainment: A Biological Assessment*. Academic Press, New York, New York.

Secretary's Commission on Pesticides and Their Relationship to Environmental Health. 1969. *Report of the Secretary's Commission on Pesticides and Their Relationship to Environmental Health, Parts I and II*. U.S. Department of Health Education and Welfare; Available from U.S. Government Printing Office, Washington, D.C. 20402.

Stern, A.C., H.C. Wohlers, R.W. Boubel, and W.P. Lowry, (Editors). 1973. *Fundamentals of Air Pollution*. Academic Press, New York, New York.

U.S. Environmental Protection Agency, (EPA). 1976. *Quality Criteria for Water*. Washington, D.C. 20460. (EPA-440/9-76-023).

Vernberg, F.J. and W.B. Vernberg, (Editors). 1974. *Pollution and Physiology of Marine Organisms*. Academic Press, New York, New York.

Wilber, C.G. 1969. *The Biological Aspects of Water Pollution*. C.C. Thomas, Springfield, Illinois.

Wolfe, D.A., (Editor). 1977. *Fate and Effects of Petroleum Hydrocarbons in Marine Ecosystems and Organisms*. Pergamon Press, New York, New York.

IMPACT ASSESSMENT AND MITIGATION

American Association of State Highway and Transportation
        Officials, (ASSHTO). 1976. *A Design Guide for
        Wildlife Protection and Conservation for Trans-
        portation Facilities*. Washington, D.C. 20045.

Arner, D.H. 1977. *Transmission Line Rights-of-Way
        Management*. Office of Biological Services,
        Fish and Wildlife Service, U.S. Department of
        the Interior, Ann Arbor, Michigan.

Bendix, S. and H.R. Graham. 1978. *Environmental
        Assessment: Approaching Maturity*. Ann Arbor
        Science Publishers, Ann Arbor, Michigan.

Cairns, J., Jr., (Editor). 1980. *The Recovery Process
        in Damaged Ecosystems*. Ann Arbor Science
        Publishers, Ann Arbor, Michigan.

Canter, L. 1977. *Environmental Impact Assessment*.
        McGraw-Hill Book Company, New York, New York.

Cheremisinoff, P.N. and A.C. Morresi. 1977. *Environ-
        mental Assessment and Impact Statement Handbook*.
        Ann Arbor Science Publishers, Ann Arbor,
        Michigan.

Council on Environmental Quality, (CEQ). 1970.
        *Environmental Quality: The First Annual Report
        of the Council on Environmental Quality*. U.S.
        Government Printing Office, Washington, D.C. 20402.
        (And subsequent reports issued annually).

Council on Environmental Quality, (CEQ). *102 Monitor*.
        U.S. Government Printing Office, Washington,
        D.C. 20402. (The *102 Monitor* is a monthly publi-
        cation of the CEQ and lists all impact state-
        ments filed with the CEQ; it is distributed by the
        U.S. Government Printing Office).

Council on Environmental Quality, (CEQ), and Federal
        Council for Science and Technology. December
        1974. *The Role of Ecology in the Federal
        Government*. U.S. Government Printing Office,
        Washington, D.C. 20402.

Council on Environmental Quality, (CEQ). 1978.
National Environmental Policy Act, Implementation
of Procedural Provisions; Final Regulations.
Federal Register, Vol. 43, No. 230, Wednesday,
November 29, 1978, 55978-56007.

Cross, F. L., Jr., and S. M. Hennigan. 1973. *Primer on
Environmental Impact Statements.* Technomic
Publishing Company, Westport, Connecticut.

Edington, J.M. and M.A. Edington. 1978. *Ecology and
Environmental Planning.* Wiley-Interscience,
Somerset, New Jersey.

Environmental Protection Agency/Corps of Engineers
Technical Committee on Criteria for Dredged and
Fill Material. July 1977. *Ecological Evaluation
of Proposed Discharge of Dredged Material into
Ocean Waters.* Environmental Effects Laboratory,
U.S. Army Engineer Waterways Experiment Station,
Vicksburg, Mississippi.

Erickson, P.A. 1979. *Environmental Impact Assessment:
Principles and Applications.* Academic Press,
New York, New York.

Erickson, P.A. and G. Camougis. 1980. *Highways and
Wetlands: Volume I, Interim Procedural Guidelines.*
Federal Highway Administration, U.S. Department
of Transportation, Washington, D.C. 20590.

Erickson, P.A. and G. Camougis. 1980. *Highways and
Wetlands: Volume III, Annotated Bibliography.*
Federal Highway Administration, U.S. Department
of Transportation, Washington, D.C. 20590.

Erickson, P.A., G. Camougis, and E.J. Robbins. 1978.
*Highways and Ecology: Impact Assessment and
Mitigation.* Federal Highway Administration,
U.S. Department of Transportation, Washington,
D.C. 20590.

Erickson, P.A., N.H. Miner, and G. Camougis. 1980.
*Highways and Wetlands: Volume II, Impact Assess-
ment, Mitigation and Enhancement Measures.*
Federal Highway Administration, U.S. Department
of Transportation, Washington, D.C. 20590.

Heer, J. E., Jr., and D.J. Hagerty. 1977. *Environmental Assessments and Statements*. Van Nostrand Reinhold Company, New York, New York.

Jain, R. K., L. V. Urban, and G. S. Stacey. 1977. *Handbook of Environmental Impact Analysis*. Van Nostrand Reinhold Company, New York, New York.

Lee, S.S., and S. Sengupta, (Editors). May 1977. *Proceedings of the Conference on Waste Heat Management and Utilization*. University of Miami, Miami, Florida.

Leopold, L.B., F.E. Clarke, B.B. Hanshaw, and J.R. Balsley. 1971. *A Procedure for Evaluating Environmental Impact*. Geological Survey Circular 645, U.S. Geological Survey, Washington, D.C. 20242.

McHarg, I.L. 1969. *Design with Nature*. Natural History Press, Garden City, New York.

National Audubon Society. 1974. *Wildlife Habitat Improvement*. Nature Center Planning Division, National Audubon Center, New York, New York.

New England Research, Inc. 1976. *Environmental Considerations*. Prepared for American Right of Way Association, Inc., 3727 West Sixth Street, Los Angeles, California 90020.

Pantell, R.H. 1976. *Techniques in Environmental Systems Analysis*. Wiley-Interscience, Somerset, New Jersey.

Rosen, S.J. 1976. *Manual for Environmental Impact Evaluation*. Prentice-Hall, Englewood Cliffs, New Jersey.

Swanson, G.A., (Technical Coordinator). 1979. *The Mitigation Symposium: A National Workshop on Mitigating Losses of Fish and Wildlife Habitats*. General Technical Report RM-65, Forest Service, U.S. Department of Agriculture, Fort Collins, Colorado.

U.S. Department of the Interior. June 1974. *The Trans-Alaska Pipeline and the Environment: A Bibliography.* Office of Library Services, Washington, D.C. 20204.

U.S. Department of the Interior, U.S. Department of Agriculture, (USDI/USDA). 1971. *Environmental Criteria for Electric Transmission Systems.* U.S. Government Printing Office, Washington, D.C. 20402.

U.S. Department of Transportation, Coast Guard. 1975. *Guide to Preparation of Environmental Analyses for Deepwater Ports.* Office of Marine Environment and Systems, Washington, D.C. 20590.

U.S. Environmental Protection Agency, (EPA). September 30, 1974. *Draft 316(a) Technical Guidance - Thermal Discharges.* Office of Water and Hazardous Materials, Washington, D.C. 20460.

U.S. Environmental Protection Agency, (EPA). April 1976. *Development Document for Best Technology Available for the Location, Design, Construction and Capacity of Cooling Water Intake Structures for Minimizing Adverse Environmental Impact.* Office of Water and Hazardous Materials, Washington, D.C. 20460. (EPA 440/1-76/015-a).

U.S. Environmental Protection Agency, (EPA). March 18, 1977. *Draft EPA Cooling Lake Assessment Manual.* Office of Federal Activities, Washington, D.C.

U.S. Fish and Wildlife Service. 1979. *Habitat Evaluation Procedures.* (Revised, March, 1979), Project Impact Evaluation Team, Division of Biological Services, Fish and Wildlife Service, U.S. Department of the Interior, Fort Collins, Colorado.

Index